Magic, Science, and Religion in Early Modern Europe

From the recovery of ancient ritual magic at the height of the Renaissance to the ignominious demise of alchemy at the dawn of the Enlightenment, Mark A. Waddell explores the rich and complex ways that premodern people made sense of their world. He describes a time when witches flew through the dark of night to feast on the flesh of unbaptized infants, magicians conversed with angels or struck pacts with demons, and astrologers cast the horoscopes of royalty. Groundbreaking discoveries changed the way that people understood the universe while, in laboratories and coffee houses, philosophers discussed how to reconcile the scientific method with the veneration of God. This engaging, illustrated new study introduces readers to the vibrant history behind the emergence of the modern world.

MARK A. WADDELL is Associate Professor at Lyman Briggs College, Michigan State University.

New Approaches to the History of Science and Medicine

This dynamic new series publishes concise but authoritative surveys on the key themes and problems in the history of science and medicine. Books in the series are written by established scholars at a level and length accessible to students and general readers, introducing and engaging major questions of historical analysis and debate.

Other Books in the Series

Barbara Hahn, *Technology in the Industrial Revolution*
John Gascoigne, *Science and the State: From the Scientific Revolution to World War II*

Magic, Science, and Religion in Early Modern Europe

MARK A. WADDELL
Michigan State University

CAMBRIDGE
UNIVERSITY PRESS

CAMBRIDGE
UNIVERSITY PRESS

University Printing House, Cambridge CB2 8BS, United Kingdom

One Liberty Plaza, 20th Floor, New York, NY 10006, USA

477 Williamstown Road, Port Melbourne, VIC 3207, Australia

314–321, 3rd Floor, Plot 3, Splendor Forum, Jasola District Centre, New Delhi – 110025, India

79 Anson Road, #06–04/06, Singapore 079906

Cambridge University Press is part of the University of Cambridge.

It furthers the University's mission by disseminating knowledge in the pursuit of education, learning, and research at the highest international levels of excellence.

www.cambridge.org
Information on this title: www.cambridge.org/9781108425285
DOI: 10.1017/9781108348232

First published 2021

A catalogue record for this publication is available from the British Library.

Library of Congress Cataloging-in-Publication Data
Names: Waddell, Mark A., author.
Title: Magic, science, and religion in early modern Europe / Mark A. Waddell, Michigan State University.
Description: Cambridge, United Kingdom ; New York, NY, USA : Cambridge University Press, [2021] | Series: New approaches to the history of science and medicine | Includes bibliographical references and index.
Identifiers: LCCN 2020037936 (print) | LCCN 2020037937 (ebook) | ISBN 9781108425285 (hardback) | ISBN 9781108441650 (paperback) | ISBN 9781108348232 (epub)
Subjects: LCSH: Magic–Europe–History. | Religion and science–Europe–History.
Classification: LCC BF1999 .W187 2021 (print) | LCC BF1999 (ebook) | DDC 940.2/1–dc23
LC record available at https://lccn.loc.gov/2020037936
LC ebook record available at https://lccn.loc.gov/2020037937

ISBN 978-1-108-42528-5 Hardback
ISBN 978-1-108-44165-0 Paperback

For Maggie

Contents

List of Figures   *page* viii

Acknowledgments   x

Introduction   1

1   Hermeticism, the Cabala, and the Search for Ancient Wisdom   13

2   Witchcraft and Demonology   44

3   Magic, Medicine, and the Microcosm   75

4   A New Cosmos: Copernicus, Galileo, and the Motion of
    the Earth   102

5   Looking for God in the Cosmic Machine   135

6   Manipulating Nature: Experiment and Alchemy in the
    Scientific Revolution   161

7   A New World? The Dawn of the Enlightenment   189

    Conclusion   203

Bibliographical Essays   209

Index   217

Figures

1.1 Portrait of Marsilio Ficino by Francesco
Allegrini, 1762.   *page* 14
1.2 The Roman Empire at its height in the second
century CE.   18
1.3 Engraving of Hermes or Mercurius Trismegistus from
Pierre Mussard, *Historia Deorum fatidicorum*, 1675.   26
1.4 A seventeenth-century depiction of the *Arbor
Cabalistica*.   34
1.5 Title page from *The Tragicall History of the Life
and Death of Doctor Faustus* by Christopher
Marlowe, 1636.   38
1.6 Portrait of John Dee, c. 1580.   39
1.7 Dee's Hieroglyphic Monad.   41
2.1 Sixteenth-century German engraving of witches being
burned alive.   46
2.2 "The Witches" by Hans Baldung Grien, 1510.   54
2.3 The Devil addressing a gathering of witches from
Francesco Maria Guazzo, *Compendium
Maleficarum*, 1626.   64
2.4 Witches performing the *osculum infame* or
shameful kiss.   66
2.5 The Devil directing witches to trample on
the cross.   67
2.6 "The Sabbath," a nineteenth-century depiction of an
early modern witches' sabbath, 1849.   68
3.1 Sixteenth-century engraving of a barber-surgeon
removing a tooth.   79
3.2 A physician examining a urine flask, from an
1849 reproduction of a fifteenth-century
engraving.   80

3.3 Title page from *De humani corporis fabrica* by
    Andreas Vesalius, 1543.   82
3.4 Man as microcosm, from Robert Fludd, *Utriusque*
    cosmi ... historia, c. 1617.   84
3.5 Zodiac signs and their astrological associations with
    different parts of the human body, from Joannes
    Ketham, *Fasciculus Medicinae*, 1495.   87
3.6 A sixteenth-century portrait of Theophrastus
    Bombastus von Hohenheim, or Paracelsus.   90
4.1 A depiction of the geocentric cosmos from Peter Apian,
    Cosmographia, 1539.   106
4.2 A 1643 woodcut depicting epicycles and deferents.   110
4.3 Portrait of Nicolaus Copernicus, c. 1515.   112
4.4 A sixteenth-century depiction of the Copernican
    cosmos with the Sun at its center.   114
4.5 Johannes Kepler's model of the solar system from his
    Mysterium Cosmographicum, 1596.   121
4.6 Engraving of Galileo Galilei, c. 1640.   123
4.7 Galileo's sketches of the surface of the Moon from his
    Sidereus nuncius, 1610.   126
5.1 Portrait of Pierre Gassendi, 1658.   139
5.2 Portrait of René Descartes, c. 1630.   143
5.3 The Cartesian universe from Descartes's
    Epistolae, 1668.   145
5.4 The mechanical transmission of sensation, from
    Descartes's *Opera Philosophica*, 1692.   152
6.1 A nineteenth-century engraving of Francis Bacon.   165
6.2 George Vertue's 1739 engraving of Robert Boyle.   172
6.3 "The Alchemist" by Philipp Galle, 1558.   176
6.4 The alchemist in his laboratory, from Michael Maier,
    Tripus Aureus, 1618.   179
6.5 An engraving from Michael Maier's *Atalanta*
    fugiens, 1618.   181
6.6 Portrait of Isaac Newton, 1689.   184

Acknowledgments

Maggie Osler introduced me to the history of ideas in an undergraduate course at the University of Calgary called, "Magic, Science, and Religion." More than twenty years later, I wrote this book in an effort to fascinate and inspire others as she did me. I think she would approve.

The students in my LB 327A course at Michigan State University waded through an earlier draft of this book as we discussed the foundations of modern science in the fall of 2018. I'm grateful to them for helping me improve it in many ways.

My thanks also go to Lucy Rhymer and Emily Sharp at Cambridge University Press for their invaluable assistance, and to the anonymous reviewers who provided thoughtful feedback at various stages of this project.

Finally, and always, I thank Matt for his patience, love, and support.

Introduction

Imagine a time when priests and inquisitors speculated about the sexual activity of demons, alchemists toiled in soot-stained laboratories to transform lead into gold, and witches flew through the dark of night to participate in wild orgies and feast on the flesh of unbaptized infants. Quack healers on every street corner promised magical cures to a credulous public while, in palaces across Europe, astrologers cast the horoscopes of queens and princes. Philosophers studied the fundamental nature of reality and, at the same time, searched for traces of God in a universe that grew larger and more complex all the time. This is the world at the heart of this book, one filled with the strange, the bizarre, the frightening, and the sacred. You might not realize that magicians played a role in creating the scientific method taught in classrooms today, that occultists laid the foundations for evidence-based medicine, or that religion was central to the study of nature at the height of the "Scientific Revolution." Yet, all of these things are true.

This book explores how the European worldview evolved between the years 1400 and 1750, and does so through a focus on three sets of beliefs and practices: *magic, science,* and *religion.* Together they defined how premodern Europeans understood their world, from the smallest of mundane events to earth-shattering cataclysms and miracles wrought by God. In many respects these realms were not distinct or separate things; for premodern people they overlapped with one another all the time, in ways that might surprise us. Eventually, however, they drifted further and further apart as the known universe expanded both literally and figuratively, leading to the worldview many of us hold today.

The Europe we will examine here existed in a precarious balance between tradition and innovation, anchored to the distant past but also pushing forward in search of unfamiliar worlds and new ideas. After 1400, that balance was threatened by unprecedented levels of

1

upheaval, conflict, and discovery, leading to a series of profound and irrevocable changes that gave rise to what many call the Western world. At the beginning of our story, the Catholic Church oversaw a unified Christian faith, science was rooted in the philosophies of the ancient Greeks, and most people believed that magic was real. By 1750, philosophers thought of the universe as a vast machine that operated with mathematical precision, the Protestant Reformation had long since splintered Christianity into different sects, and magic was mocked by the educated elite as delusions of the ignorant. This shift, according to many historians, marked the birth of the modern world. Yet, the story in this book is not one of "progress." There are no descriptions of ignorant premodern people figuring out how the universe *really* works, no tales in which the rational and logical sciences triumph over superstition and ignorance and lead everyone into a glorious future with cars and smartphones and electricity. Instead, this book is an attempt to meet historical people on their own terms. We will try to put ourselves in their shoes, however imperfectly, and understand how their beliefs about the world changed across almost 400 years.

It is important, however, that I acknowledge a profound limitation to what this study can accomplish. Because this book is essentially a history of ideas, we will be focusing most of our attention on those ideas that became particularly influential, usually as a result of their being recorded and disseminated in writing. Our narrative centers around those individuals who were able to produce, preserve, and share their ideas: in other words, members of Europe's educated elite, which really means educated men. With some notable exceptions – for example, our study of the European witch hunts – we will not examine the lives and experiences of premodern women. Likewise, we will learn relatively little about men who were uneducated or otherwise unable to participate in learned culture. Our focus will remain fixed on a tiny slice of European society, a privileged minority who held a disproportionate amount of power and influence and whose ideas, in turn, carried a disproportionate amount of weight. In some respects, these ideas did not affect the average European at all; they were part of an ongoing conversation among the elite members of society that had little relevance to the everyday problems and experiences of most people. In other respects, however, these ideas shaped European society in profound ways that affected almost everyone, sooner or later – including you.

Take a moment to consider the world around you. Unless you are reading this book in the middle of an isolated forest, chances are good that right now your body is bathed in electromagnetic fields, Wi-Fi signals, and the radiation given off by your smartphone as it communicates with the nearest cell tower. As you read this sentence, uncountable numbers of subatomic particles are passing through you, many of them hurtling from outer space. Gravity prevents you from flying off the Earth as it spins at terrifying speeds through the cosmic void, while also acting simultaneously on every other piece of matter in a universe that is incomprehensibly vast. To most of us, this is perfectly ordinary. It is simply the way things are. We accept, virtually without question, that our world is filled with unseen forces and strange phenomena like quantum entanglement and dark energy. Even if most of us don't understand what these are or how they work, we rarely stop to think about them.

In this respect, we are much like the people we will examine in this book. They, too, lived in a world full of unseen and mysterious forces. To them, it was natural and ordinary that the planets they saw in the sky influenced events here on Earth, or that some people could transform lead into gold, or that witches called on demons to help them conjure terrible storms and fatal diseases. Given how premodern people understood their universe, these were rational and sensible explanations for the phenomena that they observed around them. While today we might use different words and theories to describe our universe, for most of us it remains as mysterious and strange as it did to people living a thousand years ago.

This highlights one of the most powerful lessons that history can teach us: that human beings are connected in deep and enduring ways. Even across vast amounts of time, we are, each of us, far more alike than we are different. Another universal similarity is the human drive to ponder some of the biggest and most difficult questions: "Where did everything come from? Why does it exist? What is its purpose? What is *my* purpose?" Throughout human history there have been different systems and philosophies within which people have sought answers to these questions, including the three at the heart of this book: science, religion, and magic. Before going on, let's consider each of these in turn.

Science, as we understand the word now, is a modern invention. Its careful methodology, its well-defined disciplines, its culture of white coats and laboratories full of sophisticated technology go back perhaps 150 years; in fact, the word "scientist" was coined only in the late nineteenth century. While I have used the word "science" in the title of this book, in the chapters that follow we will focus instead on *natural philosophy*, which was concerned mainly with attempts to understand the natural or physical world. Natural philosophers considered questions like "What makes the heavens move?" or "Why do objects fall downwards?" or "How do plants and animals grow?" Among all the different kinds of knowledge pursued by premodern Europeans, natural philosophy was important but not necessarily preeminent – that honor usually went to *theology*, the study of religion and the divine. This reflects the fact that, while questions considered by natural philosophy usually started with phenomena that one could observe in the world, answers might move beyond the realm of the natural or physical and into the supernatural or the metaphysical. That is one important difference between natural philosophy and modern science, which does not accept supernatural explanations for observed phenomena. Another difference is that, while modern science relies on a single basic methodology that everyone uses, European natural philosophy did not embrace anything approaching a universal method until the eighteenth century. This explains why premodern natural philosophers often arrived at radically different answers to the same question, as we will see in later chapters.

For the most part, natural philosophy was an academic enterprise. It was carried out by the educated classes, whose members shared ideas with one another in settings like Plato's famous Academy in ancient Athens, in dense tomes written in Latin or Greek or Arabic, or in the universities that sprang up across Europe in the twelfth and thirteenth centuries. By contrast, religion and magic were both practiced by a much broader group of people in Europe. Until the nineteenth and twentieth centuries, most Europeans, regardless of class or income or education, were religious people. They believed in the existence of a supernatural power that created the universe and everything in it, and they were encouraged to live their lives according to moral and ethical principles derived from religious texts, particularly the Christian Bible. Learned and educated people might have approached some religious ideas differently than the average person, but by and large they all believed in the same God. The predominant religion in premodern Europe was Christianity, though throughout its long history there have always been different interpretations of what "Christianity" means. One important example of these differing interpretations is the Protestant Reformation, which began early in the sixteenth century and led to a permanent split between what became known as the Catholic or Roman Church and various Protestant denominations such as Lutheranism, Anglicanism, and Calvinism.

Importantly, religion had its own role in explaining things that people observed in the world. A question like "Why do objects fall downwards?" could have many possible answers, but if someone believes in a supernatural being who created the universe their answer might be simply, "Because God made things so that they fall downwards." This isn't a very *satisfying* answer, though, because it doesn't help us understand the means whereby things fall. What are the actual forces at work? Do they work the same way everywhere? Can we measure them? Virtually every single premodern European person accepted that the ultimate, final explanation for something that happened in the world was that God made it that way, but much of natural philosophy attempted to answer the more immediate questions about how and why the universe behaved the way that it did. To use the terminology that many premodern scholars employed, God was assumed to be the *final* cause for everything that happened in the universe, but natural philosophers studied the *proximate* or *immediate* causes. A natural philosopher might argue that gravity is the immediate cause for why objects fall downward, while at the same time believing that God created gravity and made it work in this way. Both "gravity" and "God" are answers to the question "Why do objects fall?", but one – gravity – is closer to us (the proximate or immediate cause) while the other – God – is much further away (the final cause). To complicate things further, premodern people accepted that God exists *above* nature; this is what the word "supernatural" means. As a result, most educated Europeans believed that God existed beyond natural laws and was free to interfere with or suspend those laws at any time. But if this is true, are they really "laws" in the first place? Could someone ever know with certainty that a particular phenomenon they observe in the world is the result of natural processes rather than of supernatural interference? The fact that religion and natural philosophy often proposed different answers to the same questions made understanding the premodern world a complicated endeavor.

Magic, as a concept, is even more difficult to pin down than either natural philosophy or religion. For the first two, we have modern parallels and analogues that, while imperfect, at least give us a foothold as we try to understand what they were in the past. But in modern societies, "magic" has many different meanings, some of them contradictory. Many people today use the term "magic" to describe trickery, fakery, and illusion, as in stage magic where hapless assistants are sawn in half only to reappear, totally unharmed. The audience knows that what they are watching is a trick, and the thrill lies in wondering how the magician did it. At the same time, others use the word "magic" to describe the actual, physical manipulation of the world, such as in works of fiction where the flick of a wand conjures up fire or water or butterbeer, or in real-world traditions like Wicca, whose followers believe that magic can bring good luck, attract wealth, or heal the sick. Someone alive today might explain magic using science and logic – for example, revealing acts of stage magic as optical illusions and clever misdirects – but for others, magic is part of spiritual systems as disparate as Louisiana Voodoo, Nordic shamanism, and LaVeyan Satanism.

In premodern Europe, magic, like religion, was not confined solely to the learned or educated classes. Large numbers of people practiced magic, from soothsayers and healers living in tiny rural communities to university professors and Christian priests who believed they could summon demons or speak with the dead. According to most educated people, magic had a simple definition: the manipulation of hidden forces to produce specific effects. Those forces existed throughout the natural world, and so the task of the magician was to study nature in order to uncover and utilize these forces as part of what they called "natural magic." For this reason, natural magic and natural philosophy often went hand-in-hand; many natural philosophers believed in the reality of natural magic, and some even practiced it. If philosophy was the means whereby one understood nature, magic was the means whereby one put nature to work.

Not all magic was natural, however. The desperate or the foolish might call on demons to help them manipulate the world and produce whatever effect they desired. Because of their supernatural knowledge of Creation, demons knew better than any human how to manipulate its hidden forces, and many premodern Europeans believed that they sometimes did so – for a price. The learned magician might have the means to summon and control a demon (the story of Faustus, which we examine in the next chapter, is about one such figure) but for the average person, the possibility of demonic interference posed a far greater threat. From the fourteenth century onward, Europe was gripped by a fever of witch-hunting, a cultural phenomenon during which hundreds of thousands of people were suspected or accused of consorting with demonic powers in order to harm their neighbors and wreak havoc on society. The magic practiced by these supposed witches was demonic in origin and truly wicked in its effects, and religious leaders, theologians, and natural philosophers all stood on the front lines of what they saw as a war waged by the forces of evil against God and His people. Magic, then, was not merely the subject of philosophical inquiry; it was of deep concern to religious institutions as well.

These three realms – natural philosophy, religion, and magic – overlapped with and affected one another in complex ways. None of them existed in isolation; together they gave shape and meaning to the world inhabited by premodern European people. Let's turn now to a very quick description of that world as it existed around the year 1400.

At this point in history the European economy was still focused around agriculture, with most people living and working in rural areas. For many hundreds of years the predominant structure of European society had been defined by *feudalism*, a system of governance and obligation that revolved around the ownership of land. A monarch granted land to members of the aristocracy, and in return demanded fealty and loyalty as well as support in times of war or hardship. These landowners then allowed other people to live on and work their land in return for taxes or other forms of support. As a result, those who worked the land rarely owned it. These people were known as *peasants*, and they made up roughly 60 percent of Europe's population at the beginning of the fifteenth century. This highly stratified and hierarchical system was one of the defining features of medieval Europe.

While a majority of Europeans lived in the countryside in the fourteenth and fifteenth centuries, however, the number of people living in towns and cities also increased steadily during this same period. The founding of the earliest European universities – the first in Bologna in 1088 CE, then Paris around 1150 CE and Oxford in 1167 CE – led to the rapid proliferation of other universities across Europe in the twelfth and thirteenth centuries. Once established, they attracted large

numbers of students to their cities while producing increasingly educated and literate populations. Alongside the rise of skilled trades and
an increase in both the production and trade of goods, the spread of
universities helped create an urban populace that could aspire to
greater wealth, education, and financial security than was possible
for most people living in rural villages.

Around 1350 CE, European society experienced one of the most
destructive and destabilizing events in its history: the Black Death, a
plague that killed as many as 200 million people in Europe and Asia by
the year 1353. Historians believe that Europe's entire population was
reduced by anywhere between 30 percent and 50 percent in just five
years, and the global population did not recover to the levels seen before
the Black Death until some three hundred years later. Crowded urban
centers were hit especially hard by plague, and as their populations
dwindled increasing numbers of people moved away from rural areas
and into cities and towns in search of work. This caused population
surges in cities across Europe, leading to rapid social and economic
changes as the urban workforce diversified into a variety of skilled
trades, the cost of manufactured goods went up, and communities of
merchants and tradesmen grew in size and wealth. Rural areas that
experienced a drop in population, due first to plague and then to the
widespread migration into towns and cities, experienced a scarcity of
agricultural workers. This led to serious food shortages across Europe.
Agricultural land became less valuable because it had fewer people to
work it, while labor became more valuable, meaning people could
demand higher wages or other forms of compensation. This created a
situation that actually favored many who continued to live and work in
rural areas. Generally speaking, agricultural workers gained more
power in relation to the aristocratic landowners, and eventually this
spelled the end of feudalism in most parts of Europe. This in turn led
to better living and working conditions for at least some people.

Our narrative, then, begins in the aftermath of the Black Death. Its
effects on European society persisted for generations, and at different
points in this book we will encounter echoes and traces of those effects:
the food shortages and economic uncertainty that contributed to the
witch hunts in the fifteenth and sixteenth centuries; the idealism of
Renaissance thinkers determined to rebuild a devastated Europe in the
image of classical antiquity; and the questioning and challenging of
authority that emerged in response to political and social instability.

There are four major themes that will be explored in this book:

1. *The influence of classical antiquity.* Ancient Greece and Rome cast very long shadows across premodern Europe. Attempts to recover classical philosophies, beliefs, and ideas played a central role in defining the cultural movement known as the Renaissance in the fourteenth and fifteenth centuries. By the sixteenth century, however, increasing numbers of educated people sought to correct or dismiss the misconceptions held by ancient writers. These efforts only accelerated in the seventeenth and eighteenth centuries, which were characterized by widespread and deliberate efforts to discard classical philosophies. Whether embracing or rejecting classical antiquity, however, there is no question that the ancient world had a profound effect on European intellectual life.

2. *The relationship between God and nature.* Throughout the history of Christian Europe, the study of nature was tied inextricably to questions about God. Most Europeans believed that the natural world represented an important means of understanding God as Creator; some even referred to the physical universe as the Book of Nature, a metaphorical text that contained crucial knowledge about the divine. Before the eighteenth century most people would have found it unthinkable to separate God from nature, which means that "scientific" inquiry often had religious implications.

3. *The problem of occult or hidden causes.* Premodern Europeans lived in a world in which there might be several possible causes behind the phenomena they observed, ranging from natural forces or properties to demons, angels, and even God Himself. Determining which causes operated in a given situation was not an idle concern, however. Failing to recognize demonic ploys could result in eternal damnation, and attributing particular effects to Nature rather than to God might lead to skepticism or atheism. This is one reason why beliefs and practices that depended on the deliberate manipulation of hidden forces, such as alchemy or magic, inspired suspicion and distrust.

4. *The interconnectedness of the premodern world.* Most Europeans, whether educated or not, saw themselves as part of a small but richly interconnected universe in which the individual was a reflection of both the world around themselves and its Creator. To study one of them was to study the others, and knowing more about the

world also meant knowing more about oneself. This interconnectedness is one reason why people invested significant meaning in the
answers to even the smallest questions.

The rest of this book explores these themes in different ways. We will
proceed chronologically, starting at the height of the Renaissance in the
fifteenth century and ending with the Enlightenment in the eighteenth
century. Each chapter presents different facets of the complex relationship between magic, natural philosophy, and religion, though the
examples and individuals we will encounter are by no means the only
ones worth knowing. Instead, each chapter presents a snapshot of a
much larger picture, all of them connected by the themes at the heart of
this narrative.

Chapter 1 explores the realm of "learned magic," a term coined by
historians to describe a set of magical traditions and philosophies with
roots in the philosophies of classical antiquity. Some of these traditions, like *hermeticism*, were first encountered by Renaissance scholars
trying to recover traces of the "golden age" of ancient Greece and
Rome. Others, like the Judaic tradition of the *Kabbalah*, had already
existed in Europe for hundreds of years but received closer attention
from Christian writers and philosophers in the fifteenth and sixteenth
centuries as they searched for new and more powerful ways of understanding their universe.

In Chapter 2 we contrast the practices of "learned magic" with the
brutal witch hunts that spread across Europe for hundreds of years,
events driven in part by fears and misunderstandings about popular or
folk magic. Tens of thousands of men, women, and children were
killed as part of the hunts, which mark a profound turning point in
European ideas about magic. This also represents an important
moment in European religious history, solidifying the power and pre-
rogative of the religious authorities to define ideas and beliefs as
heretical, or violating the teachings of Christian orthodoxy.

Chapter 3 also touches on magic, but this time in the context of
medicine. We examine how premodern Europeans practiced medicine,
including the many ways in which the wider universe was believed to
affect the human body. Physicians and other medical practitioners
commonly resorted to astrology in order to diagnose and treat their
patients, and the fundamental idea of magic – the belief that the world
contains hidden forces and powers that can be harnessed to accomplish

specific tasks – was seen as a powerful tool in the arsenal of some medical practitioners. One such practitioner was the infamous medical reformer Paracelsus (1493–1541) who, in the early decades of the sixteenth century, combined a respect for nature's secrets with a deep reverence for God in his efforts to create an entirely new way of healing.

The central role of religion in European life is a major part of Chapter 4, which explores how a small number of philosophers and mathematicians changed how people understood the cosmos. The long-standing idea of an Earth-centered universe was challenged first by the groundbreaking work of Nicolaus Copernicus (1473–1543) in the sixteenth century and then by the fiery arguments of Galileo Galilei (1564–1642) some sixty years later. After claiming that the Earth moved, Galileo was placed on trial by the Catholic Church, but while this moment in history is often used to illustrate the incompatibility of science and religion, we take a closer look at the debate in an attempt to understand the motivations and anxieties driving both sides.

Galileo's trial had a profound impact on how early modern philosophers talked about God, and in Chapter 5 we see how this affected the *mechanical philosophies of nature*, which attempted to explain natural phenomena as the motion of tiny pieces of matter. The mechanical philosophies first took shape in the first half of the seventeenth century, and they presented the educated classes with serious questions about the presence and role of God in the physical world. We follow two different mechanical philosophers, Pierre Gassendi (1592–1655) and René Descartes (1596–1650), as they sought to create systematic and useful ways of studying nature while also preserving a role for divine activity. In doing so, they created new ways of thinking that inspired later generations of scholars but also raised deep and uncomfortable questions about God, the human soul, and life itself.

Chapter 6 tackles the problem of experience, which was a central part of the "new science" that emerged in the latter half of the seventeenth century. Descending from the mechanical philosophies of Gassendi and Descartes as well as the methodology pioneered by the English philosopher Francis Bacon (1561–1626), the "new science" had as its earliest advocates virtuosi like Robert Boyle (1627–91), John Locke (1632–1704), Gottfried Leibniz (1646–1716), and Isaac Newton (1642–1727). Empiricism and experimentation both became important elements of this new approach to the study of nature, but these practices also posed particular difficulties for early modern thinkers. We examine those difficulties by looking more carefully at the theory and practice of alchemy, which was an experimental "science" in its own right but which slowly underwent a profound and lasting change in the early eighteenth century. Some practitioners of the "new science" eventually went to great lengths to separate problematic alchemical ideas from what they defined as the emerging science of "chemistry."

The slow demise of alchemy as a respectable field of study leads us into Chapter 7, which takes stock of the changed world in which European people now found themselves. This was the dawn of the Enlightenment, the sweeping cultural and intellectual movement that many historians see as the beginning of the modern West. Enlightenment thinkers called for the radical separation of religion from public life, championed rationality, and heaped scorn on the superstitions of the uneducated. The careful, sometimes precarious balance between science, magic, and religion that had survived for centuries collapsed into something that is more familiar to many of us today, with consequences that are worth considering.

1 | Hermeticism, the Cabala, and the Search for Ancient Wisdom

Around 1460, the philosopher Marsilio Ficino (1433–99) received a message from his patron, Cosimo de' Medici (1389–1464), the most powerful man in the Italian city-state of Florence. Up to this point Ficino had been hard at work translating the works of the ancient philosopher Plato (c. 424–c. 348 BCE) from their original Greek into Latin, but his patron had other ideas. He wanted Ficino to begin translating a different Greek manuscript, one that Cosimo had only recently acquired. Obligingly, Ficino set Plato aside and turned his attention to this new work. He soon realized that he had stumbled across something very important (Figure 1.1).

The works that Ficino translated became known as the *Corpus Hermeticum*, and they contained the recorded wisdom of a mysterious figure known as Hermes Trismegistus or Hermes "the Thrice-Powerful," a contemporary of Moses and a sage of unparalleled learning who had lived thousands of years earlier in ancient Egypt. His writings promised to reveal the secrets of the universe to those willing to learn, and this soon included Ficino, who became a passionate advocate for the ideas of Hermes and was instrumental in disseminating them throughout Renaissance society. Ficino, along with many others, believed that the Hermetic writings contained traces of ancient, uncorrupted wisdom that might restore human understanding to the heights achieved by those, long ago, who had known God and His creation in ways since lost to modern people.

The tradition disseminated by Ficino is known as *hermeticism*, and it incorporated both philosophical lessons on the nature of the divine as well as hands-on instructions for magical work. Both hermeticism and the other tradition we explore in this chapter, *cabalism*, are examples of learned magic – that is, magic studied and practiced by the educated elite. This is very different from the magic worked by healers, midwives, and others in small communities and rural areas across Europe, practices usually labeled by historians as "folk magic." Learned magic

MARSILIO   FICINO
SOMMO   FILOSOFO,
E CANONICO FIORENTINO.

nato a XIX Oct.bre MCCCCXXXIII.   morto il 5.no Ott.bre MCDLXXXIX.
Al merito sing.re del Rev.mo Sig.r Ant.o Gaspero Franchi Dott. in Sacra Teologia
Prot. Ap. dell' Imperial. Basilica di S. Lorenzo Canonico.
Preso da un Quadro in Ato dell' Imperial Galleria di Firenze.
Giuliano Traballesi del.   *Franc. Allegrini del.*

Figure 1.1 Portrait of Marsilio Ficino by Francesco Allegrini, 1762.
Photo by DeAgostini/Getty Imagess

had its roots in the distant past, and those who embraced it did so with
the hope that they would uncover secrets and mysteries that would
transform European society forever. This idea was so powerful and
compelling that it fundamentally altered intellectual life in Europe and
continues to inspire people today.

In this chapter we examine both hermeticism and cabalism in an
attempt to understand why they captivated European people for hundreds of years, before turning our attention to the figure of the *magus*,
the learned magician who seeks to pull back the veil that obscures the
workings of the universe. Some, like the English magus John Dee
(1527–1608), built their reputations and identities around their
reputed ability to uncover nature's secrets, but there was also widespread anxiety about the lengths to which these magicians might go in
pursuit of knowledge. That anxiety was perhaps most memorably
expressed in the fictional tale of Faustus, who trafficked with demonic
forces and paid the ultimate price. In fact, both Dee and Faustus came
to unfortunate ends, and both are useful in placing learned magic in a
wider context.

Learned Magic before Ficino

Marsilio Ficino was far from the first philosopher to write approvingly about the varied uses of magic. The practice of learned or scholarly magic existed centuries before the translation of the *Corpus Hermeticum*, most commonly among members of the clergy. The historian Richard Kieckhefer has made a close study of a fifteenth-century manuscript that he calls the "Munich handbook," a learned treatise that describes how to summon demons and work such varied magic as inducing love in women and becoming invisible. Because this text was written in Latin and demonstrated a good understanding of Christian liturgy and ritual, Kieckhefer believes that its author was a priest or monk – few others would have the theological and intellectual sophistication to create the complex rituals and ceremonies described in the handbook. In fact, because members of the clergy generally were educated in the Middle Ages, Kieckhefer argues that they were the most likely practitioners of learned magic, including magic that employed demons. The magical tradition of *necromancy*, which in classical antiquity had referred to speaking with the dead, was reimagined in the twelfth and thirteenth centuries as magic that trafficked exclusively in the summoning of demons. Thus, combining rituals of Christian exorcism and astral or astrological magic inherited from Arabic writers, medieval necromancers were well-educated men who supposedly summoned demons in order to accomplish a wide range of tasks. The Munich handbook studied by Kieckhefer is a clear descendent of this tradition.

Not all learned magic was necessarily demonic, however. Beginning in the twelfth century, Europe experienced an influx of ideas and texts from the Islamic world as part of a lengthy process of cultural exchange. This influx helped to create the first universities in Europe and set the stage for the arrival of the Renaissance, which we will explore shortly. Another consequence, however, was a shift in learned attitudes toward magic. Prior to the twelfth century, almost all theologians and philosophers believed that magic involved demonic intervention of some kind. The magical practices described in ancient literature, such as the sorceress Circe transforming men into animals in Homer's *Odyssey*, or the marvels and miracles ascribed to pre-Christian deities were understood by medieval Christians to be morally wrong and demonically inspired. Beginning in the twelfth century, however, the works of influential Islamic philosophers presented a different perspective to European scholars. Some of these authors wrote approvingly of magical practices that had nothing to do with demons and that operated by means of natural forces alone. This included the astral or celestial magic – which drew upon the hidden forces and virtues of the planets and stars – that was eventually incorporated into medieval necromancy as well as a wide range of other practices and traditions that sought to harness nature's unseen power in acts of natural magic. At the height of the Middle Ages, in the thirteenth and early fourteenth centuries, significant numbers of educated men devoted considerable time and energy to the study of natural magic, sometimes over the objections of contemporaries who still believed all magic to be fundamentally demonic in origin.

This was the backdrop against which Ficino's recovery of hermeticism took place. The Hermetic Corpus, along with a series of other works also attributed to Hermes Trismegistus, described a way of understanding the cosmos that combined religion, philosophy, and magical practice into a coherent whole. For Ficino and other proponents of hermeticism, there was nothing morally suspect about these practices; they were resolutely natural. Not everyone agreed, however, and practitioners of learned magic in the Renaissance had to contend with the same suspicions of demonic collaboration faced by learned magicians in the Middle Ages.

The Renaissance

Ficino's efforts to translate the Hermetic Corpus, and Ficino himself, were both products of the Renaissance. More particularly, Ficino's life and works exemplify the intellectual movement known as *humanism*, which sought to recover and revive the "golden age" of classical antiquity. Together, both the Renaissance and humanism set the stage for Western modernity as we know it today.

Speaking broadly, the Renaissance was a sweeping cultural movement that began in the fourteenth century and changed European society in profound and lasting ways. Because it took hold at different times across Europe, historians disagree as to when the Renaissance ended, but it persisted at least into the sixteenth century when Europe entered what is known now as the early modern period. The word "renaissance" comes from the French term for "rebirth," which leads to an obvious question: What exactly was reborn? The answer provides us with a crucial insight into premodern Europe, because what the educated classes were so eager to revive was the culture of classical antiquity. This means the culture of the ancient Greeks that existed during what is known as the *Hellenistic period*, beginning around the year 300 BCE, and of their successors, the Romans, who established the Roman Empire around the year 27 BCE and absorbed huge amounts of Greek learning, art, and culture over the following centuries.

The Renaissance, then, was an attempt to restore to Europe the art, ideals, and learning of the ancient Greeks and Romans. This fascination with the distant past defined not just the Renaissance itself, but the entirety of learned culture in premodern Europe as well. From the fifteenth century onward, the educated classes of European society devoted themselves first to the recovery of antiquity and, later, to efforts to improve upon or escape from it.

It makes sense that the Roman Empire was a powerful source of attraction and myth for premodern people, many of whom in the fifteenth century lived among tangible remnants of that once-glorious past – everything from the crumbling infrastructure of roads and aqueducts to the dilapidated ruins of a thousand settlements. At its height, around the year 120 CE, the Roman Empire stretched across a vast area and contained about 20 percent of the world's population, around 90 million people. To the west, it reached into the British Isles and what is now Spain and Portugal; to the south it covered much of northern Africa, including Egypt; and to the east it encompassed a large portion of the modern Middle East, including Mesopotamia (modern-day Iraq and Syria) and what is now Israel. It was a diverse and multicultural empire, and among its many achievements was a robust and vibrant tradition of philosophical thought inherited from the ancient Greeks (Figure 1.2).

In time, however, political instability and the sheer size of the Empire made it impossible to govern effectively, and around the year 330 CE it was effectively divided into two when the emperor Constantine the Great (272–337) moved the capital of the Empire to the city of Byzantium and renamed it Constantinople (known today as Istanbul, the capital of Turkey). Still governed from the city of Rome, the Western Empire slowly declined, a process hastened by the brutal sack of Rome itself by the barbarian hordes of the Visigoths in

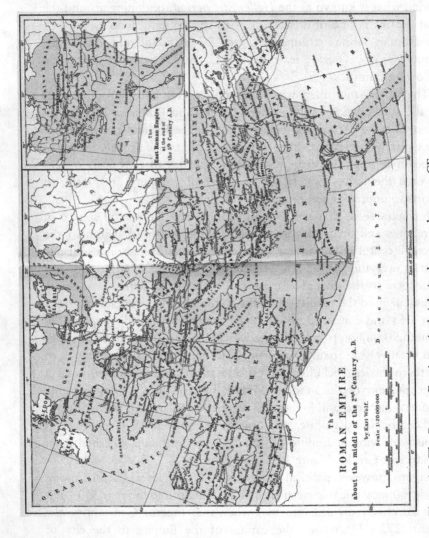

Figure 1.2 The Roman Empire at its height in the second century CE.
Photo by The Print Collector/Print Collector/Getty Images

410 CE and then the Vandals in 455 CE. By the year 495 CE, the Western Empire had fallen into ruin. By contrast, the Eastern Empire (also called the Byzantine Empire) flourished for another thousand years.

It was the Eastern Empire that safeguarded the remnants of ancient learning, most of which had been written first in Greek. This was also the language of government, religion, and learning in the Byzantine empire, as well as the language spoken daily by millions of imperial citizens, and it was in Byzantine libraries and monasteries that ancient Greek ideas were copied and preserved. Meanwhile, for people living in the ruins of the Western Empire, basic literacy in Greek had died out in what historians in the nineteenth century grimly referred to as the "Dark Ages," the period that began with the sack of Rome in the fifth century and ended more than 500 years later with the social and intellectual flourishing of the Middle Ages. For many people living in western Europe, the learning of the ancients might as well have disappeared forever.

Beginning in the seventh century CE, the Byzantines went to war with the forces of the Rashidun Caliphate, the first Islamic state founded after the death of the prophet Muhammad (c. 570–632). The Caliphate extended across much of what is now the Middle East before it was succeeded by the Umayyad Caliphate (founded in 661 CE), which expanded its territories to encompass more than 60 million people. The Caliphate warred on and off with the Byzantines but tolerated the presence of Christians and Jews within its borders. Despite the tensions between the two empires, people created both formal and informal avenues of communication and commerce. Merchants, scholars, and government officials on each side learned the language of the other, and so when the philosophical and medical works of classical antiquity eventually made their way into the Caliphate, Islamic scholars fluent in Greek began to translate them into Arabic.

The movement of classical European learning into the Islamic world sparked a period of intellectual flourishing that lasted for centuries. Under the leadership of the Abassid dynasty, which took control of the Caliphate from the Umayyads around the year 750 CE, a vast translation movement began. The House of Wisdom, or *Bayt al-Hikmah*, was established in Baghdad in the early ninth century and became arguably the most important site of learning in the entire world. Scholars translated many works of classical antiquity into Arabic and then disseminated them throughout the Islamic world. Islamic philosophers, naturalists, and physicians appropriated Greek learning and then proceeded to build their own important and original contributions on those foundations. Ibn Rushd (1126–98), known in Europe as Averroes, was an expert in Islamic law who also wrote treatises on mathematics, astronomy, medicine, and philosophy, while the ingenious Ibn-Sīnā (c. 980–1037), called Avicenna in Europe, produced hundreds of works that included some of the most important medical treatises in history, many of which were still taught in European universities in the sixteenth century. They were preceded by al-Kindī (c. 801–73), sometimes known in the Latin West as Alkindus, who was one of the first to champion the study and adaptation of Greek and Hellenistic ideas and was later revered as one of the greatest Islamic philosophers. Other prominent Islamic thinkers who emerged during this "Golden Age" include the influential physician al-Rāzī or Rhazes (854–925), al-Farabi (c. 872–950), who wrote commentaries on the philosophies of Plato and Aristotle, and al-Bīrūnī (973–1048), who was a mathematician and geographer. Alongside many others, these scholars built upon and added to Greek philosophy and medicine across hundreds of years.

But how does any of this matter to the Latin West, or to the Renaissance? The answer to that lies in the eventual reach of the Islamic Caliphate. In the eighth century, the Umayyad Caliphate expanded into the Iberian peninsula – what is now Spain and Portugal – and thereby came into direct contact with western Europe. Trade sprung up between these two cultures, and trade of course requires that each side understand the language of the other. When some in western Europe became fluent in Arabic, they were able – at last! – to read the translated works of Greek antiquity. Beginning in the eleventh and twelfth centuries, many of these works were translated from Arabic into Latin (the primary language of learning and scholarship in the west, one of the lingering remnants of the Roman Empire), until finally, after hundreds of years and a very circuitous route, the writings of the ancient Greeks finally returned to Europe. Fluency in Greek became possible again, leading western scholars to seek out Greek manuscripts still held in the east, and eventually we find ourselves in fifteenth-century Florence with the mysterious Greek manuscript that Cosimo sent to Ficino.

The recovery of ancient learning not only laid the foundations for the Renaissance but also encouraged the rise of universities, which emerged in the eleventh and twelfth centuries as centers for the translation and exchange of ideas. In fact, with the benefit of hindsight we can appreciate how thoroughly the study of classical society led to innovations that remain with us today: styles of architecture that are still visible in cities around the world; artistic practices like the use of linear perspective which transformed European art; changes to European language and education that led to the formation of distinct social classes; and new ways of studying both the natural world and the human body. Ultimately, the recovery of ancient learning altered the very fabric of European life as profoundly as it had transformed the Islamic world.

Humanism

At its most basic, humanism was the study of classical antiquity, but in fact it was far more ambitious than that. Humanists in the fifteenth and sixteenth centuries wanted to *revive* classical antiquity, not just study it. They believed that it was not enough to read and learn the ideas of classical antiquity – it was also necessary to live according to the moral philosophies and precepts passed down by classical authors. For many humanists, a virtuous life (or as many ancient philosophers put it, a life lived well) went hand-in-hand with the ability to convey one's thoughts and opinions with eloquence and clarity. They also believed that a citizenry that was educated and articulate was the ideal environment in which to revive the social and political ideals that had flourished in the classical world. Accordingly, humanism focused enormous attention on education and particularly on the subjects that had formed the *trivium*, the most basic level of education in classical antiquity: grammar, rhetoric, and logic.

Given special attention by Renaissance humanists were the disciplines of grammar and rhetoric, which were greatly expanded as part of the humanist curriculum and which became known as the *studia humanitatis*, the predecessor to what we today call the humanities. The study of logic gradually fell away, at least within this new curriculum of study, which was then expanded by the addition of history and philosophy, particularly moral philosophy. Poetry also became a subject of intense study as it was seen as the natural extension of rhetoric, the goal of which was to refine one's speech and writing so as to persuade or move an audience. Accordingly, the classical figure most revered by early humanists was the Roman orator Cicero (106–43 BCE), famed during his lifetime for his eloquence and his mastery of the Latin language.

Renaissance humanism had its roots in Italy, and many important Italian cities became cultural centers for the rise of humanism and the wider movement of the Renaissance. It is no accident that when we study Renaissance art, architecture, and literature today, we usually focus first on the works and ideas of the Italian people. Some of the most important early humanists were Italian, including the scholar often referred to as the "father of humanism," Francesco Petrarca (1304–74, widely known today as Petrarch), and they included not just philosophers but also poets and writers. As humanism spread outward from Italy other famous humanists emerged, most prominently the Dutch thinker Erasmus (1466–1536), sometimes called the "prince of the humanists."

Humanism soon integrated itself into the mindset of Europe's educated classes. Humanist educations became common among Europe's elite, and with this new style of education came a range of novel and sometimes controversial perspectives. For one thing, most of the classical authors read and idolized by Renaissance humanists had lived before the appearance of Christianity and so were regarded as *pagans*, a general term that came to mean "pre-Christian." Some, like Plato and his followers, held ideas that could be reconciled fairly easily with Christianity – for example, Neoplatonists who lived in the early centuries of the Common Era such as Plotinus (c. 204–70) and Iamblichus (c. 245–c. 325) wrote about the divine in terms that made sense to later Christians. But many classical authors held ideas that were different from, or even incompatible with Christianity, and the Catholic Church was understandably suspicious of an education that revolved around pagan or pre-Christian ideas. Furthermore, some classical authors – Cicero chief among them – had written against tyranny and praised republicanism, a system in which citizens effectively control their own state. For many early humanists, living in European monarchies that were certainly not democratic republics, the more egalitarian political vision supported by these classical authors was very appealing. This led to rising political tensions as humanist ideals spread across Europe, with long-lived consequences. There are clear connections between the rise of humanism in the fifteenth and sixteenth centuries and the modern idea of the state that began to emerge during the Enlightenment of the eighteenth century, leading to the political upheavals of the French Revolution and the American War of Independence.

As the word "humanism" implies, its primary focus was people. Modern historians generally see humanism as fundamentally concerned with the rights, freedoms, and expression of the individual, a focus still embraced today by the humanities. At the same time, however, the humanist tradition encouraged not just the revival of classical ideas but also their critique. Humanist scholars did not just blindly copy the beliefs and practices of their classical ancestors; instead, they studied, questioned, and sometimes disagreed with classical ideas. This aspect of humanism will become a central part of later chapters in this book, when we meet individuals who wanted to reform and improve upon the practices of antiquity. This interesting identity, caught somewhere between a desire to revive classical antiquity and an eagerness to reform it at the same time, exemplifies the kind of critical thinking still taught in the best traditions of humanistic education today.

The Original Wisdom

This is the complicated backdrop against which the learned traditions of classical magic emerged during the Renaissance. The ideas recovered by humanists and other scholars were not focused entirely on poetry or architecture or rhetoric; they also included ancient philosophies that ranged from theories about the natural world to expressions of spiritual and religious doctrine. Both of the traditions we examine in this chapter, hermeticism and cabalism, rest somewhere between natural philosophy and religious belief, partaking of both in different ways. This is one reason why they were so attractive to premodern people. They offered wide-ranging and comprehensive ways of understanding the whole universe, from the mundane phenomena of the everyday world all the way up to God Himself.

Even more significant was the belief that, because these philosophical traditions had ancient roots, they were closer to the original knowledge that humans had once possessed but which had degenerated after humanity's fall from divine grace when Adam and Eve were punished by God with expulsion from the Garden. Before that expulsion, Biblical scripture records that Adam had named the birds and beasts and fish, a sign to premodern Europeans that he had access to some fundamental, intrinsic knowledge that allowed him to know the true names of things. Whatever mysterious knowledge God had granted Adam and Eve, however, had been withdrawn from them after their transgression in the Garden, leaving their descendants ignorant of the deepest truths about Creation.

This knowledge is known as *prisca sapientia*, which means "ancient wisdom" in Latin. Most educated people in premodern Europe believed that there was some original wisdom or knowledge available to our earliest ancestors that had been diminished and distorted over the intervening millennia. One of the most important tasks of the philosopher was to recover fragments of that knowledge, and one way to do that was to seek out the teachings of those who lived in ancient times. These older teachings lay closer to the *prisca sapientia* and were probably less corrupted than the knowledge available in the present. This, in a nutshell, is why the figure of Hermes Trismegistus was so appealing to early modern Europeans. It is also why some Christians became fascinated by the Kabbalah, a form of wisdom passed down through many generations of Jewish mystics and scholars. Both the writings of Hermes and the Kabbalah promised to reveal to Europeans the lost secrets of a distant age.

The idea of an ancient and uncorrupted wisdom was a powerful motivation for many thinkers in premodern Europe, but it was not just a knowledge of philosophy and the physical world that had degenerated with time. Even more important was knowledge of God and the divine, known as theology. According to scripture, Adam had known God in a way that later people could not. He had spoken directly to God in the Garden, and many believed that his understanding of God was as pure and deep as it was possible for human knowledge to be. This kind of understanding was known as *prisca theologia*. Like the *prisca sapientia*, it represented a kind of knowledge that was undiminished and uncorrupted, a strain of theology that predated all later religions and therefore unified them. Adam had lost his special connection to God when he was expelled from the Garden, however, and from that moment onward human understanding of the divine had faded, corrupted by error and ignorance. The notion of an ancient theology, more closely connected to that primordial understanding of God, was extremely appealing to premodern people, and only became more so following the religious upheaval and fractures caused by the Protestant Reformation in the early sixteenth century. As with *prisca sapientia*, the *prisca theologia* had the potential to radically transform how Europeans understood their world.

Modern society tends to be forward-looking rather than fixated on the distant past, so this interest in ancient knowledge might seem strange to us now. Many premodern Europeans, however, looked around themselves and saw the tattered and faded remnants of a glorious past, one that they wanted desperately to recover. Imagine living in fifteenth-century Rome, among the crumbling ruins of the Empire, or in a remote monastery surrounded by hand-copied texts that were only fragments of a vast body of knowledge long since lost. How could you not hunger to return to a time when philosophy, art, architecture, and theology had been more sophisticated and innovative than anything that existed a thousand years later? To many educated people in premodern Europe, their society seemed only a shadow of the glories of antiquity. That perspective would eventually change but, during the Renaissance, figures like Hermes Trismegistus, whose understanding of God and nature seemed to outstrip anything available in the fifteenth century, were the key not only to recovering the past but to building the future as well.

Hermes and the Hermetic Corpus

When Marsilio Ficino recovered and translated the Hermetic Corpus from Greek into Latin, he did so as part of a larger humanist project intent on reviving classical traditions. He also established the Florentine Academy in his native Florence as an attempt to emulate Plato's famous Academy in Athens, and he was the first to translate and publish the entirety of Plato's existing works as well as the works of important Neoplatonists such as Iamblichus and Plotinus, philosophical descendants of Plato who lived in the early centuries of the Common Era. In hermeticism, however, Ficino believed he had discovered a kind of knowledge that was even more important than the writings of classical antiquity.

Recall that Ficino and many others believed that Hermes Trismegistus had been a contemporary of Moses, the prophet who was believed to have lived some 3,000 years before the advent of the Renaissance. This made Hermes an extremely valuable source of

Figure 1.3 Engraving of Hermes or Mercurius Trismegistus. From Pierre Mussard, *Historia Deorum fatidicorum*, 1675.
Photo by Time Life Pictures/Mansell/The LIFE Picture Collection via Getty Images

knowledge for premodern Europeans. He would have had access to both *prisca sapientia* and *prisca theologia* on par with that of Moses himself, and his writings that had survived down into the fifteenth century, when Ficino translated them, represented a potential goldmine of knowledge that had escaped the corruption of time (Figure 1.3).

References to Hermes, or to shadowy figures who *might* have been Hermes, exist in writings going back to classical antiquity. In the earliest centuries of the Common Era some authors speculated at length about his origins, pulling together rumors and myths to suggest that Hermes was ancient even by the standards of their time. These early writings claim that he was worshipped as some kind of combination between the Greek god Hermes – the messenger of the gods but also the patron of communication and travel – and Thoth, the ancient Egyptian god of knowledge and writing. During the Hellenistic period (roughly 300 BCE to 30 BCE), when Greece controlled much of modern-day Egypt, it was not uncommon for gods from different pantheons to become associated with one another, a practice that persisted into the later Roman occupation of the Hellenistic world.

The figure of Hermes Trismegistus appears to have arisen from this kind of overlapping association, acquiring a reputation linking him to wisdom, knowledge, and magic. While he may have been worshipped as a minor deity in antiquity, however, most Europeans in the Renaissance portrayed him as human, probably an ancient Egyptian priest or sage whose name had become associated with pagan worship.

Importantly, the Egyptian origins of Hermes intersected with a wider European fascination with ancient Egypt that persisted between the fifteenth and seventeenth centuries. During this period, numerous Egyptian obelisks covered in hieroglyphs were discovered and excavated from the ruins of ancient Rome. These artifacts had been carried from Egypt to Rome at the height of the Roman Empire, which controlled most of modern-day Egypt, but after the Empire collapsed the obelisks suffered the same fate as most of the ancient buildings in Rome, which is to say they fell over and were buried beneath a new, more modern Rome in the centuries that followed. As they were uncovered, these obelisks were then transported (with great difficulty) to various spots around the modern city and erected once more. Many are still standing today, including one situated in the center of the plaza in front of St. Peter's Basilica in the Vatican, the spiritual and political home of the Catholic Church. The enormous expense and labor required to move and restore these remnants of ancient Egypt demonstrates, at the very least, how powerfully Egyptian culture influenced the imaginations of many during the Renaissance. That same influence undoubtedly helped to make hermeticism a topic of intense interest as well.

Following Ficino's publication of the Hermetic Corpus, hermeticism grew and spread through European intellectual culture. Traces of hermetic ideas and doctrines appeared in works written by dozens, if not hundreds of authors across more than 200 years. But there is an intriguing twist. In 1614, the French scholar Isaac Casaubon (1559–1614) determined that the Hermetic Corpus had not been written at the time of Moses at all. Some of its texts had obvious Hellenistic influences, meaning that they were almost certainly written after the Greeks came to control Egypt and other parts of North Africa and the Near East. That would date these texts to sometime after roughly 300 BCE, much closer to modern times than the empires of ancient Egypt, which had flourished thousands of years earlier. Casaubon was a scholar of the classical world as well as a philologist, or someone who studies the structure of languages, and he found other clues in the hermetic texts that pointed to an origin that was even more recent than the Hellenistic period, something closer to the second or third century CE, meaning that they were compiled during the Roman Empire and after the advent of Christianity. While these texts were still ancient by the standards of early modern Europe, they were no more ancient than the Roman Empire and certainly not the product of a person living and writing millennia earlier than that.

Following Casaubon's revelation, hermeticism's influence on European thinking started to wane. Some people, however, remained utterly committed to the hermetic worldview and its philosophy, arguing that Casaubon had simply gotten his dates wrong. Today, most scholars agree with Casaubon that the Hermetic Corpus had its origins in the early centuries of the Common Era and emerged from a turbulent, chaotic mixture of Greek and Hellenistic philosophy, early Christian mysticism, and magical traditions borrowed from different sects in what is now the Middle East. These writings represent the blending of ideas – also known as *syncretism* – that was common in this period, when people eagerly mashed together the ideas and worldviews of different cultures to create entirely new ways of thinking. Maybe this mixture of so many disparate philosophies, beliefs, and practices is why hermeticism has inspired and intrigued people for hundreds of years.

The Substance of Hermeticism

The question remains, what did people find when they read the Hermetic Corpus that Ficino translated and published? First, it is important to understand that hermeticism was an extremely broad set of beliefs and ideas that combined religion, philosophy, and ancient ideas about magic. In fact, hermeticism does a superb job of demonstrating how these three systems overlapped with one another. If magic is the manipulation of the universe's hidden forces, then the hermeticist needs philosophy to investigate the universe and reveal those forces, and in doing so comes to understand how and why God created the universe in the way that He did. The true hermetic practitioner, or magus, understands that the cosmos is full of connections and invisible links between different objects, and their job is to use those connections to produce specific effects in the world. In doing so, they come closer to understanding God as Creator, who not only fashioned those connections but also uses them Himself, along with angels, demons, and the Devil. Indeed, as we will see, the figure of the hermetic magus eventually became closely intertwined with the supernatural in ways that sometimes were problematic for practitioners of hermeticism; more often than not, they were feared as the dupes or collaborators of demonic forces.

Modern scholars have made a distinction between two different kinds of hermetic writings. One group of texts, those first translated by Ficino in the fifteenth century, are more philosophical and theological in their focus. Ficino called this collection of fourteen books or chapters the *Pimander*, a term still used today to describe this relatively limited piece of hermetic writings. The *Pimander* describes the religious worldview of a sect or community that flourished in classical antiquity, and also lays out the philosophical foundations of that worldview. These texts are deeply pious, referring continually to a single God and building a philosophy of nature and life founded on the worship of this deity. This is one reason why these particular texts were so appealing to Europeans living in the Renaissance. Though everyone assumed these were pagan works, written long before the advent of Christianity, they were actually easy to reconcile with a Christian worldview because their religious elements appeared to parallel the basic tenets of Christianity. The pagan Egyptian sage Hermes revered a deity very similar to the Christian God, not only making Hermes a "good" pagan by European standards but also seeming to confirm the truly ancient roots of the Christian faith, since the hermetic texts were thought to embody an understanding of the divine, the *prisca theologia*, that stretched back to the earliest years of humankind. Now, of course, we understand that the Hermetic Corpus aligns so well with Christianity for at least two main reasons: because it was written after Christianity had appeared in what is now the Middle East, not before; and because its authors were inspired by some of the same philosophies and beliefs that also inspired the earliest Christian writers.

While there are references to astrology in the texts that Ficino translated, there is virtually nothing about magic. Instead, the works included in the *Pimander* are primarily theoretical and philosophical treatises; they describe ideas rather than their direct application. This is less true of another group of texts that began to circulate in Europe after Ficino's original translation appeared. These works are steeped in occultism and include numerous references to alchemy – the manipulation and transmutation of matter – as well as to forms of magic including divination and astrological magic. This second set of texts mentions Hermes and so are thought of as "hermetic," but they were probably written later than the theological texts, or by different people. Their emphasis is more practical; they describe how to accomplish specific tasks using the knowledge and wisdom passed down from Hermes.

There is still much that we do not know about the origins of the Hermetic writings. For one thing, these texts are extremely old and survive only in fragments of the originals or in references to destroyed or missing texts mentioned by authors of other works. For another, the Hermetic Corpus was the product of many people writing across several centuries. There is no one or universal perspective because there was no single author, or even a small group of authors. Instead, we have a sprawling collection of ideas that are connected only by references to Hermes as well as by some shared theological and philosophical arguments. If we accept this rough division of Hermetic texts into theoretical or philosophical on the one hand and practical or magical on the other, then we effectively have two different schools of thought that overlapped in some small but important ways.

One important point of overlap is the idea of salvation and empowerment through knowledge of the divine, which the historian Brian Copenhaver has described as part of his modern translation of the *Hermetica*. The mysticism present in these writings encourages the reader to contemplate the divine and, through that contemplation, comes to understand the world. This idea was itself appealing to many premodern Europeans because it encouraged a pious and religious approach to the study of nature and to philosophy more generally. But this idea also colored the kind of magic that came to be associated with hermeticism in the sixteenth and seventeenth centuries. It was a magic that was both philosophical and theological in scope: it, too, encouraged the contemplation of the divine as a form of empowerment that the magus could then use to create change in the physical world. For some, the Hermetic magus ascended a metaphorical ladder that stretched between God and the world; the higher he climbed, the more he understood and the more he was able to do with the aid of magic.

This idea of the magus ascending toward a divine understanding of the universe made sense to many people. It suggested that a pious and thoughtful person potentially could do amazing things once they grasped how everything fit together. At the same time, however, it is not difficult to see how this kind of thinking might have made some people uneasy. The suggestion that someone could use pagan knowledge to better understand the mind of God struck some Christians as a problematic, even blasphemous idea. The writings of some hermeticists also hinted that the ascended magus was capable of almost anything; in theory, understanding the hidden forces that crisscrossed the universe would give someone power that approached that of God Himself. To many, that was a dangerous proposition.

What did hermeticism look like in practice? Unfortunately, there is no one answer. Because hermeticism was so broad and represented a wealth of different and overlapping perspectives, the ways in which people adopted and disseminated it also varied widely. For example, the Italian philosopher Giovanni Pico della Mirandola (1463–94) was known to Marsilio Ficino and absorbed at least some hermetic ideas while living in Florence. He believed that there was a fundamental connection between a wide range of intellectual traditions, from the philosophy of Plato to the medieval Christian commentaries of Thomas Aquinas (1225–74) to the writings of Hermes. All of these traditions, he argued, had stumbled upon truths about the world, and so all of them, to varying degrees, shared something of the *prisca sapientia*, that first and original knowledge about Creation once granted to Adam in the Garden. Pico's willingness to place Christian philosophy alongside the writings of the pagan Greeks and even the Arab infidels led to accusations of heresy and a papal ban on any public discussion of these ideas, but his belief in a shared foundation between these different traditions was very reminiscent of Ficino's own ideas.

Over the course of the next century or so, hermeticism grew and evolved until it became a sprawling, often untidy collection of ideas, beliefs, and claims. The term "hermetic" became increasingly common in early modern texts, but those texts could, and often did, contradict one another. Hermeticism, as well as the figure of Hermes himself, came to represent very different things to different people. For example, the Venetian philosopher Francesco Patrizi (1529–97) helped to disseminate hermetic ideas to wider audiences in the generation after

Ficino, and also wrote approvingly of the close links between Platonic philosophy and Christianity evident in the Hermetic writings. Like many others who lived and wrote in the turbulent decades of the sixteenth century, during which the Protestant Reformation fractured Christianity seemingly beyond repair, Patrizi wanted to find evidence of a pure, uncorrupted theology that might heal the divisions that now separated Catholics from Protestants. For him, as for both Ficino and Pico before him, hermeticism offered a way to uncover ancient truths about God.

Robert Fludd (1574–1637), an English physician and philosopher, was heavily influenced by Hermetic ideas when he wrote about the elaborate harmonies that spanned the cosmos and the alchemical processes capable of breaking down material substances to expose their fundamental natures. Though very different from Patrizi in almost every way, Fludd was also committed to the search for theological and religious truths in the study of the natural world. He created what he called a "theophilosophy" to express this, and assigned supernatural causes to natural phenomena in ways that made many of his contemporaries uneasy – for example, he speculated that the angels themselves transmitted forces like magnetism from one object to another. Long after his death, Fludd was revered by some as one of the most important hermetic philosophers in early modern Europe, and his vision of a universe in which everything was connected to everything else appealed to many.

Given its emphasis on original wisdom and knowledge, perhaps it is not surprising that hermeticism managed to overcome at least some of the strife and division caused by the Reformation and the fracturing of Christianity. Fludd was an unapologetic Protestant who wrote freely against Catholic dogma, but there were hermeticists among Catholics as well. The Jesuit philosopher Athanasius Kircher (1602–80), living in the heart of Rome and writing with the express permission of the Catholic Church, made no secret of his admiration for Hermes. In his work on the magnet, first published in 1641, Kircher claimed that everything in the world was connected by "secret knots" and included lavish images in which metaphorical chains bound together disparate things. Tellingly, he also ended that same book by calling God "the great Magnet" whose divine love was the force that drew everything together. In some ways, Kircher's vision of God was not all that different from the divine being described by Hermes himself.

Patrizi, Fludd, and Kircher were very different people who lived and wrote in very different contexts, yet all of them were drawn to the idea that deep, fundamental truths about the universe had been recorded in the Hermetic Corpus. Whether seeking the ancient roots of Christianity or unlocking the means to manipulate nature using the unseen correspondences that connected all things, these and many other thinkers were drawn to the mix of philosophy, religion, and magic that characterized hermeticism. This tradition confirmed for many people that *prisca sapientia* and *prisca theologia* were real, and that their recovery would illuminate the truths of the universe. Hermeticism also encouraged educated Europeans to see magic as connected to the divine and, more broadly, the supernatural world; it suggested that the study of nature's hidden forces and powers could be both a spiritual endeavor and a path to tremendous power.

The Cabala

The word "cabala" usually refers to the Christianized form of the Hebrew Kabbalah, a mystical and esoteric form of knowledge with its roots in Judaism. For centuries people have used these traditions to interpret religious and mystical texts, gaining access to secret or hidden knowledge contained in these works. Jewish scholars have speculated that the origins of the Kabbalah date back to Moses receiving the Law (usually known in Christianity as the Ten Commandments) from God. When God revealed the Law to Moses, He also revealed the secret meaning of the Law which became the Judaic tradition of the Kabbalah. Its practitioners believe that the revealed word of God as recorded in the Law and also in the Hebrew Bible, or the Tanakh, contains secret and hidden wisdom that can be discovered only by the study and manipulation of these texts. This emphasis on the written word and, by extension, the letters of the Hebrew alphabet reflects a belief in the power of language as well as the idea that, for God, words had generative power – after all, scripture records that God spoke the universe into existence when He said, *Fiat lux*, or "Let there be light." There are also kabbalistic traditions that focus primarily on the ten *sefirot*, which are interpreted sometimes as manifestations or emanations of God's various attributes and, at other times, as ten different names for God. Together, the *sefirot* form the foundation of the physical universe; they are the means by which God created everything

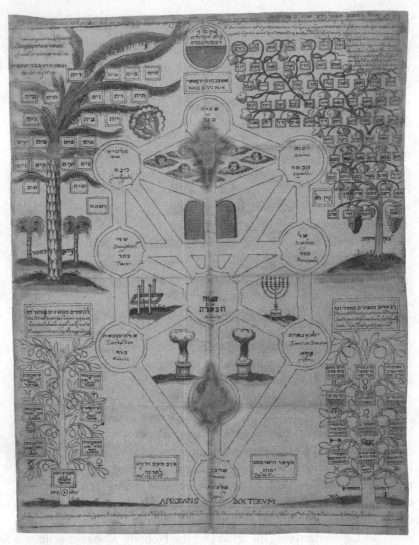

Figure 1.4 A seventeenth-century depiction of the *Arbor Cabalistica* or Cabalistic Tree, showing the ten *sefirot*.
Photo by Fine Art Images/Heritage Images/Getty Images

but also part of God Himself. By studying the *sefirot*, the kabbalist comes to understand both God and His creation (Figure 1.4).

The Judaic tradition of the Kabbalah is rich and complex, and while it should be understood primarily as a form of mysticism through which an individual seeks to know God, throughout its long history it has also embraced magical elements as well. After all, understanding the structure of the universe is the first step in learning how to manipulate it. Kabbalistic magic already existed in the Judaic tradition when it attracted the interest of Christian scholars, and it did not take long for the Christianized cabalistic tradition to incorporate magical practices as well.

Given the hostility with which many premodern Europeans viewed Jewish culture, it might seem surprising that a Jewish mystical tradition would interest Christian philosophers at all. If we remember the European fascination with both *prisca sapientia* and *prisca theologia*, however, it starts to make a little more sense. It was widely acknowledged that Judaic theology and mysticism predated the rise of Christianity, leading some to believe that Judaism contained truths that had since been lost or corrupted. Some of the earliest European interest in a Christian cabala originated with the *conversos*, individuals who had converted from Judaism to Christianity in Spain and Portugal in the fourteenth and fifteenth centuries. Many *conversos* sought to reconcile these two faiths or, in some cases, to find an effective way to convert others from Judaism to Christianity. As they sought a common ground between the two, they inspired some Christians to do the same thing.

Prominent among these Christians was Giovanni Pico della Mirandola, the protégé of Ficino, who is generally acknowledged today as the first Christian to disseminate the cabala to a wide audience. He became fascinated with both the Kabbalah and hermeticism toward the end of the fifteenth century and believed that the revelation of mysteries was possible only by combining and synthesizing the various traditions of distant antiquity into a single philosophy that included Aristotelianism, Neoplatonism, hermeticism, and the Kabbalah. He presented this unorthodox vision in his *Nine Hundred Theses*, a series of arguments that brought together natural philosophy, theology, and several different occult traditions. He intended to hold a public forum in Rome where scholars from across Europe would debate his ideas, but in 1487 Pope Innocent VIII (1432–92) intervened and ultimately declared some of Pico's ideas to be heretical, meaning that they directly contradicted Church doctrines and teachings. Pico's work became the first printed book to be banned by the Church, after which most copies were confiscated and destroyed.

Those who came after Pico, such as Johann Reuchlin (1455–1522), Francesco Giorgi (1467–1540), and Giordano Bruno (1548–1600), continued to explore the connections between the mystical Kabbalist tradition and Christian theology. Reuchlin's *De Arte Cabbalistica*, published in 1517, became especially influential in the study of the Christian cabala. Reuchlin argued that the cabala offered one of the best ways of demonstrating the truth of Christianity as well as the means to connect the study of nature with the veneration of God. Guillaume Postel (1510–81), who came of age after Reuchlin's death, used the cabala to define and understand the many connections or correspondences that he thought existed between different things. Like the hermeticist Fludd and his fellow Jesuit Kircher, Postel conceived of the universe as a vast tapestry in which fundamental truths about nature could be revealed by tracing the hidden links between disparate objects and phenomena.

Ultimately, interest in the cabala stretched into the early decades of the eighteenth century, though its popularity among the educated classes had decreased significantly by that point. Since then, both the Judaic Kabbalah and the Christianized cabala have experienced periodic revivals in the West, often coinciding with broader revivals of interest in mystical and esoteric ideas. The same is true for hermeticism, with which the cabala shares many important similarities. Both assumed prominence as part of Renaissance humanism and its recovery of ancient ideas. Both were esoteric traditions that embraced the discovery and revelation of mysteries, and both had strong connections to *prisca sapientia* and *prisca theologia*. Advocates and proponents of these traditions saw them as a means of reforming or correcting religion by returning it to its most ancient roots, as well as a means of gaining knowledge about the natural world. Both hermeticism and the cabala were unmistakably religious or spiritual endeavors, but many of their followers also believed that there was a practical dimension to the knowledge they revealed – they each promised the ability to manipulate the world in certain ways, which we should understand as the hallmark of a magical tradition. At the very least, there was widespread agreement among philosophers and scholars that the more one understood the foundations of the universe, the more easily one could manipulate them, and that understanding was central to both hermeticism and the cabala.

The Learned Magician

As interest in learned magic spread across Europe, so too did anxieties and fears about what its practitioners might accomplish. Consider, for example, two different but overlapping depictions of the learned magician or magus. One is a fictional character named Faustus or Faust, the protagonist of different stories that circulated for centuries before reaching a much wider audience in a popular play written by the English playwright Christopher Marlowe (1564–93). The other is a real person who not only lived at the same time as Marlowe, but whose life may have influenced Marlowe's depiction of Faustus: the English astrologer, mathematician, and philosopher John Dee (1527–1608). Between them, Faustus and Dee reveal why practitioners of learned magic were seen by some as frightening personifications of hubris and folly.

At its most basic, the story of Faustus demonstrated for European audiences the moral and physical dangers of learned magic. Its origins are unclear; there are cautionary tales and stories about Faustus that stretch back to the Middle Ages, many of them first circulating in what is now Germany. Some of those stories later became associated with tales of an unnamed traveling magician who supposedly lived around the beginning of the sixteenth century and who blasphemously claimed to be able to replicate the miracles of Christ through the use of magic. It is difficult to know if any of these tales were true, however, meaning that Faustus may have been a real person, a myth constructed from innuendo and imagination, or both. His story was popularized for English audiences by the playwright Christopher Marlowe, a contemporary of Shakespeare, in his *The Tragicall History of the Life and Death of Doctor Faustus*. First performed sometime around 1590, Marlowe's play helped to cement the reputation of Faustus as a learned magician who consorted with dark forces and ultimately met an untimely and gruesome end (Figure 1.5).

In the play, Faustus is a wise and learned man who, through many years of patient work, had uncovered knowledge about almost everything in the world. But he still hungers for more, and so he turns to a disreputable magician for help in summoning a demon. The idea that demons or the Devil himself could be asked to aid the desperate or the foolish has a long history, and the next chapter, which examines European witchcraft, will have much more to say on this subject. In

Figure 1.5 Title page from *The Tragicall History of the Life and Death of Doctor Faustus* by Christopher Marlowe, 1636.
Photo by DeAgostini/Getty Images

the context of the Faustus myth, however, the summoning of a demon played into a common belief concerning magic and those who practiced it – namely, that they worked with dark forces and powers in pursuit of power and mastery over the world. In Marlowe's telling, however, Faustus summons the demon Mephistophilis and strikes a bargain with Lucifer (the Devil) not because he is inherently corrupt or evil but because he is curious. He desires knowledge rather than power, at least at first. He asks for twenty-four years of life, during which he will be able to work magic; in return, he promises his soul to Lucifer. But Faustus does not receive the knowledge he wanted so desperately. The demon Mephistophilis, though supposedly his servant, refuses to answer his questions, and Faustus ultimately abuses his magical knowledge to perform cheap tricks and pranks for the next twenty-four years. Finally, as his contract with Lucifer comes due, Faustus realizes his mistake and tries to repent, but it is too late. The play ends with Faustus being dragged from the stage and into Hell, screaming to God for help.

Marlowe's play was a big hit with Elizabethan audiences. There were rumors that real devils and demons appeared on stage during the performances, drawn there by the story of the infamous Faustus, and that some members of the audience were driven mad when they

Figure 1.6 Portrait of John Dee, c. 1580.
Photo by Hulton Archive/Getty Images

saw these supernatural apparitions. Ultimately, Marlowe's retelling of the Faustus myth was concerned more with the dangers of pride and arrogance rather than those posed by learned magic, but it still provides us with some insight into how early modern people thought about magical practitioners. Among philosophers and naturalists, hermeticism, cabalism, and other forms of esoteric knowledge were valued because they offered the possibility of understanding the universe in new and powerful ways. They represented both the wisdom of antiquity and the promise of new innovations to come. But to others – members of the clergy as well as ordinary people – learned magic was dangerous, an attitude that harkens back to the medieval belief that all magic was inherently demonic. The clever wiles of demons and devils too easily could ensnare the careless or arrogant magician, and this was a danger not just to the magician himself but also to those around him. For many, magic was a threat to social order and public welfare, and it was this same threat that drove the persecution of both the learned magician and the uneducated witch.

This idea of persecution or suspicion is a key part of the story of John Dee (Figure 1.6). He was already a well-known, even infamous figure by the time Marlowe wrote his play, and it is possible that both Marlowe and his audience drew connections between the fictional character Faustus and the real-life magus Dee. The beliefs of the latter demonstrate not just the powerful attractions of the occult and esoteric knowledge we have examined in this chapter, but also the ways in which this knowledge was intertwined closely with religion and natural philosophy in this period. Dee was heavily influenced by hermeticism and by the Neoplatonic philosophy of Marsilio Ficino, and it is clear that he was also well versed in the subject of the cabala. Like Ficino and many others, Dee saw both philosophy and magic as necessary to draw back the metaphorical curtain and apprehend the truths that lay behind the physical world.

Dee rightly should be understood first and foremost as a mathematician, though in the sixteenth and seventeenth centuries the concept of "mathematics" was very broad. In Dee's case, his mathematical expertise encompassed astronomy, astrology, navigation, and mechanics, all of which we today would consider examples of applied mathematics. But Dee also believed that mathematics was the key to more esoteric knowledge as well. He claimed that numbers were the key to understanding the universe. In this, he was not very different from contemporary thinkers like Galileo Galilei, who we will encounter in Chapter 4. Galileo was also a highly skilled mathematician who argued that mathematics was the language with which God had created the world. In most other respects, though, Dee and Galileo were very different people; for example, Galileo had no interest in hermeticism or the occult philosophies that so fascinated Dee.

Dee's skill with mechanics led him to build some dazzling special effects for a play produced while he was at university, and the results were so convincing and strange that he was accused by some of trafficking with dark powers. This led to his being widely regarded as a magician during his lifetime, a label that he occasionally resented. He became an unofficial advisor to the young princess Elizabeth, and when she ascended to the English throne in 1558, the date of her coronation was determined by Dee's astrological calculations. Later in life she appointed him her court astrologer.

In 1564 Dee published the *Monas Hieroglyphica*, a work with strong hermetic as well as cabalistic influences. The Hieroglyphic Monad is a symbol that Dee invented which expressed, for him, the unity of the cosmos and its elements (Figure 1.7). Its exact meaning remains a mystery, however, because the commentary that Dee published in the *Monas Hieroglyphica* is so obscure as to be impossible to decipher. Dee's ideas had a significant influence on others, however –

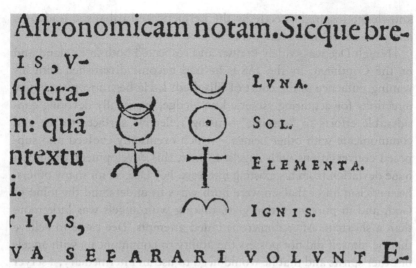

Astronomicam notam. Sicque bre-

I S , V-
sidera-
m: quã
ntextu
i.

LVNA.

SOL.

ELEMENTA.

IGNIS.

ˡIVS,

VA SEPARARI VOLVNT E-

Figure 1.7 Dee's Hieroglyphic Monad from his *Monas Hieroglyphica*, 1564.
Image courtesy of the Wellcome Collection

the Monad appeared in numerous works from this period, including an infamous treatise that claimed to describe a hidden esoteric community known as the *Rosicrucians*, or the Fraternity of the Rosy Cross. Manifestoes laying out the existence and purpose of the Rosicrucians appeared across Europe in the first decades of the seventeenth century, with one of the earliest being the *Chymical Wedding of Christian Rosenkreutz*, first published in 1616. It was no accident that the *Chymical Wedding* featured a prominent image of Dee's Hieroglyphic Monad; the manifesto and its ideas both emerged from the same tradition of esoteric knowledge in which Dee himself was immersed.

The Rosicrucian manifestos presented the recovery of ancient esoteric wisdom as the key to humanity's spiritual "reformation," and together they demonstrate how traditions such as hermeticism and cabalism evolved during the sixteenth and seventeenth centuries, long after Ficino's original translation of the *Corpus Hermeticum*. There is little evidence that the Rosicrucians actually existed, however; their manifestos may have been part of a grand hoax, or perhaps the idealistic imaginings of a single person. Even if this shadowy society did not exist, however, the philosophy laid out in the *Chymical Wedding* and other works described a quest for esoteric and occult knowledge inspired directly by the hermetic and cabalistic traditions and personified in practitioners like Dee.

Though Dee was widely known and respected both in England and on the Continent, by the 1580s he had become dissatisfied with his waning influence in the court of Elizabeth I. He began to pursue new directions for acquiring esoteric knowledge, eventually devoting considerable efforts to "scrying" – using reflective surfaces to see and communicate with other beings – which eventually evolved into supposed conversations with angels. In fact, this development was not a huge deviation from his existing interests. For Dee, as for many others, hermeticism and cabalism were both ways to understand the mind of God, and in pursuit of that goal, talking with angels was little more than a shortcut. After numerous failed attempts, Dee came to believe that he himself did not possess the ability to communicate with angels or other spirits and that he would need to use an intermediary, a scryer who could gaze into a polished mirror of obsidian or a crystal ball and both see and hear the voices of the angels. After some unsuccessful attempts with several different scryers, in 1582 Dee met Edward Kelley (1555–97), a rather mysterious figure who would eventually acquire as colorful a reputation as Dee himself. Some accounts claim that Kelley was a convicted forger who had had part of his ears removed as punishment, and historians are divided as to whether Kelley's work with Dee was sincere or an elaborate exercise in fraud and deception.

It was through Kelley that the angels spoke to Dee for many years. They dictated entire books, written in an angelic language that only Kelley could understand. Dee was seeking a remedy for the religious divisions that had fractured Christendom following the Protestant Reformation, and he turned to the angels for a pure and ancient theology – the *prisca theologia* – that he believed could unify the warring factions of Christianity. For Dee, this was first and foremost a religious endeavor; occultism and esoteric philosophies like hermeticism were merely the means of uncovering hidden religious and theological truths.

Dee and Kelley, with their wives, traveled to the Continent and met with both the Holy Roman Emperor Rudolph II (1552–1612) and Stefan Báthory (1533–86), king of Poland. Rudolph in particular was fascinated with occultism and alchemy, and Dee and Kelley had hopes of receiving his patronage, though this never materialized. They spent several years in central Europe until, in 1587, Kelley passed along a surprising message from the angels: They wanted Dee and Kelley to share all of their possessions, including their wives. Dee was troubled by this revelation and broke off his relationship with Kelley – but not before the two men swapped wives for an evening. Dee returned to England while Kelley remained in Prague and worked as an alchemist in the employ of Rudolph II. Kelley promised Rudolph that he could transmute base metals into gold, and the Emperor was so impressed that he knighted Kelley in 1590, only to imprison him the following year. The rest of Kelley's life was a mixture of opulence and hardship. He was imprisoned repeatedly by Rudolph for failing to produce the promised gold, and it is generally believed that he died after sustaining serious injuries while trying to escape from prison.

As for John Dee, he returned to England to find that his vast library had been partially destroyed by an angry mob fearful of Dee's reputation as a sorcerer. This exemplifies the fear and anxiety that so often accompanied even the suspicion of magic in this period. Dee was not the first to suffer consequences for his interest in the esoteric and occult arts, and he certainly was not the last. The fate of his library and of Dee himself – he died in poverty at the age of 82 – shows us that the learned pursuit of magic could be a dangerous thing.

2 | Witchcraft and Demonology

In 1612, in the English county of Lancashire, eleven people were accused of practicing witchcraft. They were placed on trial and key evidence against them was produced by a nine-year-old girl who was related to many of the accused. After a two-day trial, nine of the accused were hanged, including the grandmother, mother, sister, and brother of the nine-year-old witness, a girl named Jennet Device. (Jennet herself was accused of being a witch more than twenty years later, and her accuser was a ten-year-old boy.) This was the now-famous case of the Pendle witches, and it had many of the hallmarks of the witch hunts that took place across Europe for several hundred years. Most of those accused were women, and they were thought to have caused death, illness, and hardship by cursing their neighbors with magic or sorcery. Some of the Pendle women admitted to being witches, perhaps as the result of harsh interrogations or the threat of torture, but others of the accused protested their innocence all the way to the gallows.

Today, the small town of Pendle has capitalized on this grim history, using it to attract tourists and hosting a gathering of self-proclaimed witches atop Pendle Hill every Halloween. In fact, the European witch hunts continue to fascinate people all around the world. We have inherited from this period in European history the idea of a "witch hunt" as the unfair and unjust persecution of an individual or group of people, along with stereotypes of witches and witchcraft that continuously reappear in popular culture. Nor were the hunts confined to Europe alone – the infamous persecutions that took place in and around the town of Salem in colonial Massachusetts remain an important part of early American history. All of this suggests that the premodern witch hunts continue to resonate with us today. This goes beyond a ghoulish fascination with torture or the outlandish stories that so-called witches and their accusers told one another; there is something about these hunts that still speaks to us now. It might be a deep-rooted fear of persecution and accusation, or maybe the uncomfortable fact that this history shows us how quickly neighbors, friends, and families might turn on each other, either in the name of a higher good or through sheer desperation and self-preservation.

Witchcraft cuts across all three realms explored in this book. Its connection with magic is obvious. Witches were often accused of *maleficia*, which were acts of malevolent or evil magic. *Maleficia* sickened or killed people and animals, summoned destructive storms and other disasters, and otherwise caused harm. It was common for inexplicable or frightening events to be blamed on *maleficia* and, by extension, on witches. This meant that forms of local or traditional magic, already present in European communities for centuries, assumed a more complicated identity during this period. Popular beliefs and ideas about magic repeatedly came into contact with learned theories about witchcraft proposed by theologians, magistrates, and philosophers, sometimes turning what had been perceived as harmless or helpful practices into something far more dangerous.

There is also a clear religious dimension to the hunts, as almost all learned theories about witchcraft assumed that the Devil and his demons were responsible for the temptation and seduction of witches as well as for many of their harmful actions. The witch hunts emerged from a premodern worldview in which the forces of good and evil were thought to be at war with one another and conscripting ordinary people to fight their battles. During the period of the witch hunts the Devil and his demons went from being a distant, abstract threat to something much closer to home – literally, in many cases, as people began to accuse their neighbors of witchcraft.

At the same time, explorers discovered new continents filled with strange and unfamiliar things while astronomers peered into the farthest reaches of an expanding cosmos. Philosophers and naturalists began to realize how little of the world they actually understood. Witches, like the magicians we examined in the last chapter, were able to manipulate the natural world in many different ways, though their understanding of nature itself was usually limited. Nonetheless, the witch offered a way of understanding nature's hidden powers at a moment when philosophers were grappling with the discovery of entire new worlds. It made for an uneasy time, ripe for misunderstanding and fear.

Figure 2.1 Sixteenth-century German engraving of witches being burned alive.
Photo by Photo12/Universal Images Group via Getty Images

This chapter explores how all of these ideas intersected to create a new and frightening concept, that of the heretical, devil-worshipping witch who secretly tried to engineer the downfall of European society. As this idea took hold in the collective European consciousness, anxiety and fear reached epidemic proportions and ultimately led to more than 200 years of intense paranoia and persecution. It is difficult to know exactly how many people died as a result of the witch hunts because many detailed records either were not made at the time or have not survived to the present day, but scholarly estimates suggest that between 40,000 and 60,000 women, men, and children were executed as witches both in Europe itself and its colonies (Figure 2.1). Many more were accused, and many of those were tortured, exiled, or punished in various ways. The number of lives, families, and communities affected by the witch hunts is staggering.

The fear and paranoia that led to the hunts were rooted in the idea that the witch was the ultimate secret agent. He or she looked and acted just like their neighbors, often providing valuable services such as healing or midwifery and living what seemed to be an ordinary Christian life. But in reality, the witch devoted years to slowly and methodically destroying their communities by murdering the innocent, conjuring storms and other disasters to devastate livestock and crops, and luring more and more people into service to the Devil. The witch was a frightening figure, a living representation of anarchy and dissent in a Europe that was already wracked by war, religious conflict, and economic instability.

Ultimately, the witch hunts were a perfect storm. Beginning in the early decades of the fifteenth century, European society experienced profound and rapid changes: massive demographic shifts due to the staggering losses of the Black Death and subsequent migration from rural to urban centers; a rising tide of public unrest exacerbated by food shortages and the breakdown of the feudal system that had supported society for hundreds of years; and a Church trying to contain repeated outbreaks of unorthodox beliefs while dealing with its own internal tensions. All of this left European people struggling to adapt to an uncertain and unstable world, producing sustained feelings of fear, anxiety, and anger that came to be focused on the figure of the witch.

Heresy and the Fear of Rebellion

The idea of *heresy* assumed a central role in how premodern people, particularly the educated classes, understood and responded to the existence of witches in European society. Heresy is the conscious and deliberate denial of specific doctrines or ideas held by a religious institution, such as the Catholic Church. While there are examples of dissent and disagreement in the history of every faith, heresy came to mean something particular in medieval Christian Europe. The Catholic Church is widely regarded to have been the most important institution in premodern Europe, at least until the sixteenth century when the Protestant Reformation divided Christians into different sects. The Church was a constant presence in the life of virtually every European person across many centuries, which means that deliberate or organized dissent from the teachings of the Church potentially called into question the very foundations of European society. Moreover, because heretical movements were frequently inspired by thoughtful and reasoned objections to certain points of orthodox Christian doctrine rather than by simple ignorance, often they were led by people with arguments that might be persuasive or compelling to others. From the perspective of the Church, then, the real danger of heresy was its potential to spread and encourage widespread challenges to the Church's authority.

One of the most studied instances of premodern European heresy was called *Catharism*, which was centered largely in southern France between the twelfth and fourteenth centuries. Its roots are unclear, though there are many similarities with unorthodox Christian movements and philosophies that existed in the Near East, particularly the Paulician sect which thrived in and around what is now Armenia between the seventh and ninth centuries. Catharism is usually described as dualistic, meaning that it believed in two equal but opposite beings or powers, one good and one evil. While some early Christian sects, like the Paulicians, may have held similar views, by the Middle Ages such a belief was incompatible with Church doctrine, which taught that there was only one supreme being (God) who was wholly good. The primary being of evil in Christian mythology, the Devil or Satan, was not equal to God, an idea that we will revisit later in this chapter.

What we know of Cathar beliefs is fragmentary because the Church destroyed everything it could find during its attempts to obliterate this heresy. Some records do survive, however, including those produced when Cathars were placed on trial in ecclesiastical courts and questioned about their beliefs. Most of them identified the God of Good with the God of the New Testament, while they saw the God of Evil as the God of the Old Testament. Because this evil God had been responsible for the creation of the world, the Cathars saw all physical substances as inherently tainted. They rejected the sacraments and rites of the Church, which they believed to be corrupt, and effectively created their own religion. All of this would have been troubling to the medieval Church, but even more problematic was the fact that many visible and vocal Cathars were educated people with a sophisticated understanding of the Christian theology they rejected. This made them doubly dangerous, because they had the knowledge and means to sway those less informed to their heretical ways of thinking.

Ultimately, the Cathars came to a bloody end. Pope Innocent III (1161–1216) launched the Albigensian Crusade in 1209 CE, a campaign that lasted for twenty years and focused the might of the Church and the French crown on the Languedoc region of southern France. Thousands of people were killed, whether they were Cathars or not. As just one example, when the forces of the Crusade attacked the French city of Béziers, there was a question as to how they were to distinguish heretics from the innocent. In reply, the commander of the Crusade's forces supposedly said, "Kill them all and let God sort them out." As many as 20,000 people were slaughtered in Béziers, Catholics and Cathars alike.

The fate of the Cathars shows how concerned the Church was about the problem of heresy. While they all but disappeared following the Albigensian Crusade, in its wake the Roman papacy created the Inquisition, tasked with identifying and persecuting heretics. The Inquisition would eventually become a powerful force in premodern Europe, playing a significant role in the persecution of suspected witches as well as the censuring of philosophers and thinkers who dared to question the doctrines of the Catholic Church.

Folk Magic and Popular Religion

The European persecution of witchcraft had a long and complex history, stretching from the late Middle Ages through to the eighteenth century. The witch hunts were not necessarily at the forefront of European life for the whole of this period, but they persisted for a long time; entire generations of people were born, lived, and died in the shadow of the hunts. They represent an important turning point, in that this is the period in European history when certain kinds of magic ceased to be relatively harmless beliefs and became instead the cause of an international crisis.

Long before the hunts began, people who were believed to practice magic could be found in communities across Europe. Many of these individuals practiced what scholars call "folk magic" – that is, magic that was not learned or necessarily literate. This was different from traditions of magic such as the cabala or hermeticism, which were practiced almost entirely by educated men and which had deep roots in the philosophies of antiquity. Instead, folk magic was practiced by both men and women, usually as part of oral traditions in which skills and techniques were passed from one person to another by apprenticeship or hands-on learning.

The collective work of many historians, sociologists, and anthropologists suggests that folk magic exists in myriad forms across human cultures. For example, the anthropologist Bronislaw Malinowski (1884–1942) famously described the connections between magic, religion, and psychology that he observed among the indigenous peoples of the Trobriand Islands, in Papua New Guinea. His work followed in part from sociologist Émile Durkheim's (1858–1917) *The Elementary Forms of the Religious Life*, first published in 1912, which examined various forms of "primitive" religious belief and their connections to magical practices. These early studies were highly influential, and suggested that every human society has individuals who claim a special connection to forces or powers that are hidden from most people, an idea borne out in more recent scholarship.[1] These individuals might be known as shamans, priestesses, or healers, and members of their communities often believe that they can contact the gods or other powers beyond the mortal world. They might also be known as magicians, sorcerers, or witches, with access to hidden forces that allow them to do amazing and useful things. The lines between religion and magic, then, are often blurred; practitioners of both have claimed access to a hidden world and received respect because of it.

Indeed, for many people living in premodern Europe the spells of a local practitioner of folk magic and the prayers of a Christian priest may have seemed remarkably similar. In both cases, they called on forces or powers beyond the mundane world and asked that they grant specific boons or results. Both probably seemed equally effective, at least to those who believed that they would work. At the same time, both might also appear equally mysterious. The symbols scrawled by a magician in the casting of a charm were probably no less puzzling to the typical person than the words written in the Bible, given the relatively low levels of literacy in fifteenth-century Europe. The Catholic mass, conducted in Latin, was undoubtedly familiar to those who heard it every week, but the average person was no more capable of understanding Latin than they were the magical phrases uttered by a practitioner of folk magic. In fact, such magic often incorporated phrases or names from Christian ritual, further confusing the issue.

A village priest in fifteenth-century France or Germany or Spain would have been horrified by the suggestion that his Christian prayers and rituals were the same as the spells and charms produced by a local practitioner of folk magic. Nevertheless, there is ample evidence that, for the average person living in Europe, the distinctions between these two realms were sometimes very small. We know this because the

---

[1] For an overview of this scholarship, see Rebecca Stein and Philip L. Stein, *The Anthropology of Religion, Magic, and Witchcraft*, 4th ed. (New York: Routledge, 2017).

Church repeatedly tried to educate ordinary people about the differences between the two, sending priests into rural areas in an effort to
eliminate "superstitious" practices such as praying to particular
Christian saints for good luck or using the consecrated host or bread
to protect people and livestock from harm. Further complicating the
issue, some members of the clergy, particularly in smaller communities,
occasionally inspired this behavior. A local priest might have trekked
out to freshly plowed fields in spring and sprinkled them with holy
water to encourage a good crop and, more importantly, to reassure his
parishioners that God and His Church had the power to grant these
blessings. The next day, farmers may have gone to their local practitioner of folk magic for a charm to provide the same thing, perhaps
believing that if help from one person was good, help from two people
was even better.

Folk magic meant different things in different communities, but
there are some generalities worth exploring. This kind of magic was
relatively simple, especially when compared with the elaborate rituals
and philosophies of traditions like hermeticism. It often revolved
around healing, midwifery, or simple charms to bring luck or to avert
misfortune – in other words, the kinds of useful benefits that virtually
anyone would want. Its practitioners were not necessarily learned or
literate, though they might have been. At the very least, they possessed
knowledge that most others did not, however it was acquired. They
used local plants and other materials in their magic, and many of them
drew on a deep repository of local beliefs, probably going back
many generations.

Practitioners of folk magic are sometimes conflated with what were
known in Britain as *cunning folk*, in France as *devins-guérisseurs*, the
benandanti in Italy, and by many other names in other countries. We
would probably categorize these individuals as practitioners of *folk
medicine* because much of what they provided to their local communities revolved around healing the body and treating illness. In many
places, they were more likely to be women than men. They were often
respected members of those local communities, but we need to understand that not all practitioners of folk medicine used magic, not all
practitioners of folk magic were witches, and not all witches practiced
folk magic. This is important because in many cases these distinctions
were made by premodern people themselves, even at the height of the
witch hunts. Where records of witch trials exist, they often show that

people had to be convinced or persuaded by members of the educated classes to see their local purveyor of magical charms as a dangerous, Devil-worshipping witch. This idea of the witch, which was a product of the educated classes, did not mesh with how ordinary people understood and experienced folk magic in their own communities. Nevertheless, when a community began looking for witches, it was not uncommon for practitioners of folk magic to be among the first suspected by their neighbors.

The Witch

The figure of the witch changed a number of times over the course of the witch hunts. In the earliest trials for witchcraft, which began around 1420, those accused as witches were as likely to be men as women. This is probably because the earliest witch trials overlapped with contemporary trials for heresy, particularly in Italy, and heretics could be of either sex. It was also not uncommon for practitioners of ritual or ceremonial magic – almost all of whom were men – to be accused of *maleficia* in this early period. Over time, those suspected of witchcraft were accused not only of practicing harmful magic but also of trafficking with the Devil and gathering in groups to perform profane and blasphemous rites, beliefs that developed as part of a general theory of witchcraft created by the educated classes. By the late fifteenth century, as this theory was consolidated in a number of important and widely circulated texts, the crime of witchcraft assumed the general configuration that persisted for the rest of the hunts.

Beginning in the 1560s the witch hunts reached truly epidemic proportions in large parts of Europe, and a pervasive stereotype of the witch flourished between roughly 1560 and 1650, when the hunts were at their height. This was an important development in the history of European witchcraft because it focused attention on particular kinds of people who then were persecuted in disproportionate numbers. In some respects this stereotype still persists today. When you think of the stereotypical witch you probably have a few basic characteristics in mind: female, usually older, perhaps ugly or misshapen in some way, and almost always a figure of malice or evil. This is the cartoon version of the witch that appears in movies and Halloween decorations, and once you recognize it you start to see it everywhere in modern culture.

That stereotype has changed remarkably little from the idea of the witch that evolved in premodern Europe. In the late sixteenth and seventeenth centuries the typical witch was female, usually older, of limited means and with some knowledge of herbs, simple medicines, or midwifery (Figure 2.2). She might be employed as a cook or make a living by curing her neighbors of ailments. In many cases she occupied a social position on the margins of her local community, perhaps as a widow or unmarried spinster with little in the way of family or friends. This made her both vulnerable and suspect. Poverty and marginalization were seen as obvious reasons why someone might be tempted by the offers of wealth and power made by demons, who in turn demanded fealty of the would-be witch. Likewise, those on the margins of communities are sometimes mistreated by those with more power or status, something that was widely understood by premodern people. The beggar, the itinerant peddler, the widow eking out a meager living by dispensing herbs and medicines – these were precarious lives, just small steps away from abject poverty. Add the paranoia of the witch hunts and you have a situation in which misfortune suffered by one member of a community might be blamed on another, and when searching for a convenient scapegoat, social standing and old grudges often became important. As a result, those with marginal social status were usually the most vulnerable to accusations and persecution.

Of course, there were exceptions to the prevailing stereotype. Some who were accused and executed as witches were wealthy or important members of their communities, including men in positions of authority. In the diocese of Trier, in what is now western Germany, almost 400 people were executed for the crime of sorcery or witchcraft in the late sixteenth century. A witness later wrote that "the madness of the furious populace" was so fierce and uncontrolled that it led to the executions of powerful men including a judge, city councilors, burgomasters, and priests. Similarly, a former magistrate named Johannes Junius was executed for witchcraft in 1628 in the German city of Bamberg after confessing to sexual relations with a demon and to poisoning a horse. In some parts of Europe, such as Iceland and Finland, as well as in Russia, significantly more men were accused of witchcraft or *maleficia* than women, probably because in these societies men, rather than women, tended to be practitioners of folk magic. Moreover, in the largest witch hunt in European history, which took

Figure 2.2 "The Witches" by Hans Baldung Grien, 1510.
Photo by Culture Club/Getty Images

place in the Basque region of Spain between 1609 and 1611, more than 1,300 children confessed to attending large gatherings of witches while also implicating thousands of men, women, and other children. All of this demonstrates that the stereotype of the witch was not fixed or static – it varied from one place and time to another, perhaps in response to different social dynamics and pressures.

While significant numbers of men were tortured and executed during the witch hunts, however, roughly 70 percent of those accused and persecuted as witches were women. The roles and occupations most often filled by women, particularly surrounding childbirth and healing, were more frequently subject to suspicion because they were associated with events that were especially frightening for premodern people – namely, illness and death, and especially the death of children. When infants sickened or died, suspicion often fell on the women who assisted with the birth or with caring for both mother and child in the weeks afterward. After examining records of many such cases, the historian Lyndal Roper argued that many of the women suspected of being witches were initially accused by other women as part of complex psychosocial dynamics surrounding childbirth, motherhood, and infant death. Under the influence of the paranoia and fear encouraged by the witch hunts, the same knowledge and skills that made some individuals useful members of their communities could also make them seem dangerous.

In the nineteenth and twentieth centuries most historians assumed that the witch hunts primarily had been a rural phenomenon. They believed that smaller towns and villages were ripe for witch panics because their populations tended to be more superstitious and less educated than those in urban centers, and because those same populations were usually small and close-knit. It may well be the case that smaller communities, in which every individual is dependent upon and living in close proximity to his or her neighbors, tend to give rise to the simmering social tensions and grudges that can sometimes explode into public accusations of wrongdoing. Urban centers in premodern Europe, by contrast, did not usually encourage close-knit bonds between neighbors, making it less likely that social tensions would reach a dangerous tipping point.

The available evidence would seem to support this generalization. The majority of witchcraft accusations and persecutions did take place in rural communities, sometimes jumping from one small village to another as more and more people were accused. But we also know that plenty of witch panics arose in larger towns and cities as well, especially in Germany and Poland. The social dynamics might have been different from those at play in smaller communities, but urban centers were not immune to the pressures and tensions that produced witch-craft accusations. In fact, because people lived in closer contact with one another in towns and cities, witch panics tended to be larger and more concentrated, leading to greater numbers of people being accused and persecuted.

While scholars have speculated that social dynamics played an important role in triggering accusations of witchcraft, it is challenging to identify those dynamics when examining specific cases. This is largely because most of the materials that historians use to study this period – usually written records of specific trials – often lack such information. While both secular authorities and ecclesiastical institu-tions like the Spanish and Roman Inquisitions made records of witch trials, they tended to include the specific accusations made against individuals and perhaps summaries of the testimony or evidence pre-sented but rarely anything else that might help to reconstruct who these people were to their neighbors or why these particular individuals were accused in the first place. This is unsurprising if we consider that it was usually the religious and judicial elites who produced those records, and that they likely were not familiar with the local issues that usually precipitated these trials.

Even if the specific dynamics at play in particular witch hunts are difficult to reconstruct, however, it is clear that they were a crucial part of the process whereby individuals were accused as witches. The learned elite may have created the theory of witchcraft and overseen the questioning and trials of the accused, but the initial accusation was almost always made by a neighbor or family member. Moreover, where evidence has survived, it shows that these accusations happened within complex social networks that were specific to that particular community. There are cases in which those accusing someone of witchcraft referenced events that had taken place years or even decades in the past. For example, before the midwife Walpurga Hausmännin was executed in the diocese of Augsburg in 1587 she confessed to having killed forty-one infants over a period of twelve years along with a number of other acts of *maleficia* such as conjuring hailstorms to destroy crops in incidents that went back even further. Such cases suggest that long-standing grudges, disputes, and social disruptions drove people to accuse others of *maleficia* and witchcraft. This may also explain why older individuals were more likely to be accused than the young. They had lived in their communities for decades and thus may have accumulated problematic or disruptive relationships with many of their neighbors over time. Another explanation is that the elderly of both sexes were more likely to exhibit signs of dementia or mental illness, leading to behavior that their neighbors would have considered suspect and perhaps even dangerous.

If local dynamics and relationships often gave rise to accusations of witchcraft, it is also the case that the social roles typically occupied by women – healers, midwives, and caregivers for other women's children – were disproportionately targeted by witchcraft accusations. These roles supposedly provided women with opportunities to perform harmful acts of magic; a practitioner of folk medicine, with a good knowledge of local herbs and probably some folk magic as well, would be able to sicken or kill rather than heal. Likewise, the midwife was often suspected following the death of a new mother or her child, and many people also believed that witches stole infants and devoured their flesh at secret gatherings. In fact, anyone in regular contact with children could be suspected of causing illness or death, and given how many children died in their first years of life in premodern Europe, there was a lot of suspicion to go around.

Because women were more likely to occupy social roles that allowed them to harm others, particularly the vulnerable such as young children or expectant mothers, they became the obvious suspects when misfortune struck. At the same time, women themselves tended to be more vulnerable to accusations than men, particularly if they were elderly, widowed, or impoverished. They made easy targets for accusations because there were fewer people willing or able to protect them, but at the same time they were also more likely to seek employment as midwives or caretakers as a means to support themselves, creating a vicious cycle in which women were limited to particular social roles that in turn made them objects of suspicion. We must also add to this the fact that many of those accused of witchcraft were practitioners of folk medicine, folk magic, or both, and that many such practitioners were women. Such practices left people open to accusations of witchcraft, and this is at least a partial explanation for why so many women were caught up in the witch hunts. Ultimately, the fact that most of those accused of witchcraft were women is neither incidental nor coincidental. It is central to this history.

While there are significant challenges in trying to reconstruct the social dynamics present in specific premodern communities, there were also more widespread economic and social factors that played a role in the witch hunts, and fortunately these are easier to study. For example, we know that the prices of food and other goods increased steadily in the fourteenth and fifteenth centuries while economic disparity between different groups of people widened. In urban areas, merchants, skilled tradesmen, and their families enjoyed greater prosperity and access to resources while those in rural areas had to contend with higher prices and, in some places, a lack of available work. There was widespread famine in the later decades of the sixteenth century, which intensified social and economic pressures. Religious strife and warfare were also common in the sixteenth century, thanks in part to the turmoil caused by the Protestant Reformation, leading to massive disruption across much of central Europe. It is no coincidence that these upheavals accompanied the heights of the witch hunts; most historians now assume that the two were in fact closely related. The more unstable and precarious life became for many Europeans, the more they searched for explanations and, in some cases, people to blame.

Witchcraft Theory

What makes the story of European witchcraft so interesting is the way in which learned and popular cultures came together to create the idea of the witch. Folk beliefs and educated theories both fed into one another in a cycle that led to centuries of persecution. This is markedly different from the traditions of high or learned magic that we saw in the previous chapter, which were exclusively the realm of the educated elite, and yet those same learned traditions of magic helped to shape the anti-magic backlash in European culture that culminated in the witch hunts. The ritual or ceremonial magic in those traditions, which was seen originally as relatively innocent, slowly became associated with demonic or diabolical forces by philosophers and theologians concerned about the moral implications of magical meddling. We saw this in the case of John Dee, who was feared as a demon-conjuring sorcerer by his neighbors, as well as in the long-lived mythology of Faustus. It was a relatively small step from associating learned magic with demons to associating *all* magic with demons.

While it was true that local or regional beliefs sometimes had an influence on how theologians, inquisitors, and magistrates understood witchcraft, it is important to note that the *theory* of witchcraft – the body of knowledge that outlined and explained the practices of witches – was almost entirely a construction of the educated classes. Witchcraft theory was not static and unchanging; it grew over time, absorbing new ideas and adding them to what the historian Brian Levack has called the "cumulative theory" of witchcraft. In this way, beliefs that had once been separate from one another – for example, the claim that witches could fly through the air, or the belief that witches gathered in secluded places to perform terrible and blasphemous rites – eventually became part of a unified theory, until many people took it as a fact that witches flew on sticks, brooms, or animals to their secret gatherings, called *sabbats*. The cumulative nature of this kind of thinking made witchcraft theory remarkably adaptable and resilient, meaning that it was more likely to survive and grow over time.

Those who developed and disseminated the learned theory of witchcraft were theologians, philosophers, and lawyers. Prominent theorists included Johannes Nider (c. 1380–1438), whose *Formicarius* was one of the earliest works to describe witches as predominantly female; Jean Bodin (1530–96), whose *De la démonomanie des sorciers,* or "On the Demon-mania of Sorcerers" first appeared in 1580; Nicolas Remy (1530–1616), whose 1595 treatise *Demonolatreiae libri tres* was influenced heavily by Bodin; and Francesco Maria Guazzo (c. 1570–c. 1630), who published his *Compendium maleficarum* in 1608. Their ideas, as well as those of many others, influenced the beliefs of local magistrates and inquisitors who typically tried suspected witches. Without this foundation, it is unlikely that the witch hunts would have become as severe and pervasive as they did.

Witchcraft theory, then, was cumulative, integrating new ideas or perspectives over time and becoming both more complex and more difficult to challenge. It also represented the intersection of different concerns and anxieties held by the educated elite: religious and theological fears about the rise of heresy in Europe and the presumed role of the Devil in stoking these challenges to the authority of the Church; worries about magic and its corrupting effect on human morality; and the increasingly elaborate philosophical study of demonology and the Devil himself. Over time, all of these concerns coalesced around the figure of the witch.

The *Malleus Maleficarum*

The *Malleus maleficarum* or "Hammer of the Witches" was one of the most important influences on ideas about witchcraft in premodern Europe. It was first published in Germany in 1486 and went through thirteen editions over the next thirty-five years, suggesting that it was extremely popular. Its primary author was an inquisitor named Heinrich Kramer (c. 1430–1505); a second author, Jacob Sprenger (1438–95), is often mentioned as well, but his influence on the *Malleus* is unclear. In 1484, after Kramer had tried to imprison and interrogate a number of suspected witches in the German town of Ravensburg, he had run into local opposition from both secular and ecclesiastical authorities who resented his trying to put their people to the question. Kramer then persuaded Pope Innocent VIII (1432–92) to issue a papal bull or declaration, known as the *Summis desiderantes*, which formally acknowledged the existence of the heresy of witchcraft and affirmed the authority of inquisitors (like Kramer) to go about "correcting, imprisoning, punishing, and chastising" those suspected of being witches. In 1485, with his authority now affirmed by the pope, Kramer led a witch hunt in the diocese of Brixen (in what is now part of Austria and northern Italy) before publishing the *Malleus maleficarum* the following year.

Kramer wrote the *Malleus* as a kind of instruction manual for inquisitors, magistrates, and other judicial authorities. It laid out the facts of witchcraft as Kramer saw them, with particular emphasis on a couple of key ideas. First, all witches were heretics and servants of the Devil. As heretics, they fell squarely within the purview of the Church and, more specifically, its inquisitors. Witchcraft, then, was fundamentally a religious concern; it was a problem for the Church to address with the expected and unwavering support of secular authorities. Second, Kramer laid enormous emphasis on what he saw as the profound moral and spiritual weakness of women. Most witches were women because they were easier to tempt and corrupt. Their minds were weaker than those of men, making them more likely to be deceived, and their nature was closer to sin, something affirmed by the original sin committed by Eve in the Garden. While Kramer's contempt for women is notable by modern standards, for the time his ideas would have been relatively uncontroversial.

The *Malleus* also devoted considerable energy to framing witchcraft as a fundamentally sexual crime. Kramer wrote a great deal about the perverse lusts that drove witches to copulate with demons and even with the Devil himself, and argued that it was sexual desire – to which women were more prone than men – that lay at the root of the heretical crimes committed by witches. "All witchcraft," he wrote, "comes from carnal lust, which in women is insatiable." This attention to sexuality and desire was not confined to the *Malleus* alone, however; many witchcraft theorists and commentators focused on sex, and specifically on the sexual acts that supposedly occurred between witches and demons. The historian Walter Stephens has argued that this focus on sex was driven not by simple prudishness or the repressed desires of the (male, celibate) inquisitors, but rather by anxiety and doubt. If a witch confessed to physical intercourse with a demon, this could be taken as tangible and empirical evidence that demons actually exist, something that Stephens claims was by no means a certainty for many theologians and theorists. In other words, sexual relations between witches and demons may have reassured the inquisitors not just that witchcraft was real, but that demons were real as well. And if demons were real, then so was God.

The suggestion that the witch hunts may have been motivated by religious doubt is intriguing. Is it possible that centuries of persecution and tens of thousands of deaths were the result of a relatively small group of people searching desperately for proof that the supernatural existed? I believe that it is. We will see in later chapters that the search for God's existence was not confined just to the witch hunts. Philosophers and theologians turned to ever more elaborate proofs for God as the known universe grew larger and more complex, and even in the late seventeenth century some scholars saw witchcraft as one way to prove that the supernatural realm actually existed.

The Devil and the Pact

It might seem obvious that the central figure in the witch hunts was the witch, but in an important sense the real source of fear and anxiety was the Devil. In Christian mythology the Devil is a fallen angel who rebelled against the will of God and was cast down into Hell along with a number of other, lesser angels. Those angels became demons, and premodern Europeans believed that both demons and the Devil himself roamed the earth freely, tempting humans into terrible sins. Though that idea had existed for centuries, it took on new life in the sixteenth and seventeenth centuries when the witch hunts were at their height. Europeans came to believe that demons were everywhere, recruiting human agents for a war against Christendom. War, disease, political strife, social upheaval – all could be explained by the shadowy presence of diabolical forces working against the better interests of humanity. Given that Europeans experienced a great deal of war, disease, strife, and upheaval in this period, it is not surprising that they looked to explanations beyond the realm of the ordinary to make sense of a world gone mad.

The Devil and his demons were frightening because they were powerful, though theologians and philosophers agreed that their power was still limited when compared to that held by God, who literally could do anything. He was not bound by the laws of nature and could violate them whenever He wished, sometimes working miracles such as raising the dead to life or curing terrible diseases in an instant. The Devil, however, *was* constrained by the laws of nature. He could not perform real miracles, and at least in theory his ability to manipulate the world was no greater than our own. But he had an important advantage: He had once been an angel who was present when God first created the universe. He knew and understood all of the hidden forces and secret causes that lay just beneath the surface of the world, and that meant that he knew how to manipulate them. As an example of this belief, consider the words of the papal judge Paolo Grillandi (born c. 1490). In 1536, he wrote in his *Tractatus de hereticis et sortilegiis* ("Treatise on heretics and fortune-tellers") that the Devil was "the best philosopher…and the most excellent physician." Later in the sixteenth century, the theologian and demonologist Paulus Frisius (born c. 1555) claimed that the Devil was "an excellent student of physics, astronomy, and mathematics."[2]

This level of knowledge was frightening in itself, but of greater concern to premodern people was what this knowledge allowed the Devil to do. Because he and his demons could manipulate the natural world in profound and inexplicable ways, it was easy for them to trick or deceive. The Devil was thought to be a master of illusion, and this

---

[2] Stuart Clark, *Thinking With Demons: The Idea of Witchcraft in Early Modern Europe* (Oxford: Oxford University Press, 1999), p. 162.

was arguably his greatest power because he could make the average person believe almost anything. He did not need to violate natural laws in order to perform miracles if he could merely *convince* someone that they had seen a miracle. Through trickery, it was possible for the Devil to sway someone's mind to such an extent that they became his follower, believing his claims and doing his bidding. Many accused witches clung to such explanations for their supposed crimes, claiming that they had been deceived by a being with incredible powers of persuasion. This rarely saved them, however; many inquisitors argued that the truly virtuous should be able to recognize demonic trickery.

Looking back to the figure of Faustus, we can see why the incredible knowledge possessed by demons and the Devil presented an exciting opportunity to the ambitious magician: If he wanted access to that knowledge, he only had to make an agreement with a demon and it would be his. Curiosity alone might seduce the educated into striking a pact with the Devil or his demons. This is one important reason why the traditions of high magic were seen as so problematic by many people, particularly theologians: They potentially opened the unwary and the foolish into making deals with demonic forces. More frightening was the possibility that a magician might make a pact with a demon without even realizing it. This was often called an "implicit pact," an agreement between the magician and a demon that was never openly articulated but which existed all the same. This might take place when a demon secretly aided someone's efforts to practice ritual magic, but unfortunately, whether the magician was aware of the pact or not, their soul was still damned to an eternity in Hell. This was why many theologians counseled against practicing any form of magic, no matter how innocent one's motivations might be.

This possibility was particularly worrying to John Dee, the magus we encountered in the last chapter. Dee conversed with angelic beings for many years, but he was always aware of the possibility that any of those conversations might actually involve a demon in disguise rather than an angel. His diaries sometimes reflect a healthy skepticism about the supernatural beings who appeared in his scrying stone, but that skepticism was not enough to dissuade him from pursuing his angelic communications. In this respect, Dee demonstrates the fine line walked by some practitioners of learned magic, one that stretched between a desire for greater knowledge and the possibility of eternal damnation.

Figure 2.3 The Devil addressing a gathering of witches. From Francesco Maria Guazzo, *Compendium Maleficarum*, 1626.
Photo by Universal History Archive/Universal Images Group via Getty Images

If someone as educated and powerful as John Dee worried about demonic or diabolical interference, imagine how much more perilous it might be for the average person, someone who probably lacked Dee's learning and skepticism. The vast majority of those accused of witchcraft were not educated people, and ignorance made it easy for a potential witch to be duped by demons or the Devil. But what happened once someone was fooled into accepting the Devil's help? This is where we encounter one of the most important elements of premodern witchcraft theory: the diabolical pact.

As witchcraft theory evolved, it incorporated the idea that witches were assumed to have entered into a formal agreement with the Devil or one of his demons, striking with them a pact or promise (Figure 2.3). This was important to premodern witchcraft theorists for several reasons. First, it confirmed that the Devil and his demons were directly involved in the spread of witchcraft in this period. More than that, it transformed witchcraft into a campaign waged by the Devil against the forces of righteousness. This provided an obvious and powerful justification for efforts by the clergy, the ruling classes, and the educated elite to stamp out witchcraft wherever it might exist. Second, the pact confirmed that the witch was in fact a heretic, someone who knowingly and deliberately set themselves against the teachings of the Christian church and its scriptures, and as we already know, heretics were seen as enemies not just of Christianity but of European society as well.

The third reason why the pact was so important to witchcraft theorists was because it confirmed that the witch was a knowing ally of the Devil. In the case of an "implicit pact" that an ignorant magician might inadvertently make with the Devil or a demon, while that pact might ultimately damn the individual to an eternity in Hell, the fact that he did not make it knowingly or deliberately at least tipped the scales slightly in his favor. But during the witch hunts it was widely assumed that witches knew exactly what they were doing. They might have been duped by a demon at first, but ultimately they offered up their fealty and their souls to the Devil in return for power, riches, and comfort, promising in return to carry out acts of harm and wickedness against their neighbors. This moral weakness on the part of the witch tended to overlap with contemporary ideas about the general weakness of women, as exemplified in texts like the *Malleus maleficarum*. This in turn provided a means of justifying and explaining why so many accused witches were women.

In the fevered imaginations of most witchcraft theorists, the witch made his or her pact with the Devil under suitably horrifying and blasphemous circumstances. These pacts were often struck during large gatherings of witches, the sabbats. The individual would be presented to the Devil, who might appear in any number of forms – a tall, dark man, a black goat, or even a misshapen creature with both human and animal features – and who would demand particular acts of the would-be witch. A common one was the *osculum infame* or the "shameful kiss," where the witch kissed the anus of the Devil in an act of supreme degradation (Figure 2.4). Usually the individual was also required to defile or trample on symbols of the Christian faith, including the cross or crucifix and the host (the bread given to the Christian faithful as part of the Eucharist) (Figure 2.5). Once they had performed these symbolic acts, the witch was welcomed into the company of his or her fellow witches with a huge feast and, often, an orgy.

Figure 2.4 Witches performing the *osculum infame* or shameful kiss.
Photo by DeAgostini/Getty Images

An interesting note is that many of the acts that witches were
believed to commit at their sabbats looked very similar to the activities
supposedly perpetrated by other suspect or heretical groups. In the
twelfth and thirteenth centuries, for example, the Cathars often were
accused of performing blasphemous parodies of the Christian Mass.
Because they believed that procreation was sinful (it trapped an innocent soul within a corrupt and inherently evil body) they also were
accused of inducing abortions and eating the miscarried fetus in ritualistic fashion. Similarly, Jewish communities were sometimes accused of
mocking Christian rituals, murdering the children of Christians, or of
poisoning the symbols of the Christian Mass, usually the consecrated
bread or host. It is unlikely that any of these really happened, but it is
noteworthy that those who were caught up in the witch hunts assumed
the frightening and immoral characteristics of other groups that had
also provoked the collective anxiety of premodern Christians.

We might ask why premodern Europeans believed that someone
would make a diabolical pact and participate in such activities. In

Figure 2.5 The Devil directing witches to trample on the cross.
Photo by DeAgostini/Getty Images

many cases, the answer is surprisingly simple: Life was hard for just about everyone. Disease and illness were widespread in ways that we can hardly imagine today. Due to the religious and political instabilities caused by the Protestant Reformation in the early sixteenth century, conflict and warfare were continuous, dangerous problems as well; this was particularly true in the Holy Roman Empire and surrounding territories in central Europe, where neighboring principalities, city-states, and provinces were often divided by religious differences. Most people in Europe lived hand-to-mouth, scarcely making enough money to support themselves and their families, and for rural communities a single hard winter or wet summer might lead to widespread starvation and deprivation. Given these realities, it is not hard to understand why someone might leap at an offer of power and wealth and privilege, even if that offer came with some hefty strings attached. For many people in premodern Europe, the idea that their neighbors, friends, and loved ones might be tempted to follow the Devil was frighteningly plausible.

Figure 2.6 "The Sabbath," a nineteenth-century depiction of an early modern witches' sabbath, 1849.
Photo by The Print Collector/Print Collector/Getty Images

The sabbats are worth examining because they too had an important place in witchcraft theory. The idea that witches gathered secretly to worship the Devil and celebrate their wicked confederacy took hold of the European imagination in a profound way. It was believed that witches would gather from the surrounding area, many of them flying on common household implements like brooms or pitchforks, and engage in a series of stereotypical activities that included performing blasphemous perversions of the Catholic Mass, murdering and eating unbaptized infants, and sexual orgies (Figure 2.6). In many Nordic countries, witches were often suspected of using these gatherings to conjure terrible storms that could sink ships, while in parts of Spain and Italy there were tales of hundreds of people dancing with wild abandon on lonely hilltops. In England, where the witch hunts were relatively restrained and there was little anxiety about Devil worship, the sabbats were often imagined as fairly sedate affairs where witches merely shared a huge feast; few people in England believed that witches committed the horrifying and lurid acts that appeared in witch trials in central Europe.

For witchcraft theorists the sabbat demonstrated that the witch was a willing and enthusiastic participant in Devil worship, but more importantly it meant that a captured witch could identify other witches that they had seen at these gatherings. One of the most pervasive characteristics of the later witch hunts was the way in which a suspected witch, when questioned, almost always implicated other people as witches. Moreover, they usually did so under the extreme duress of judicial torture. This is why hunts grew from isolated accusations of a single individual to those involving groups of people, from the Pendle witches we met at the start of this chapter and those accused in Salem, Massachusetts all the way to the massive hunts in Trier, Bamberg, and the Basque region of Spain in which hundreds or even thousands of people were accused of witchcraft. When threatened with or subjected to torture, almost everybody pointed the finger at their family, friends, and neighbors, if only to spare themselves.

The Process of Persecution

It is not entirely clear what uneducated people believed about the role of demons in acts of malicious magic before the witch hunts began, but we do know that throughout the hunts, people were generally more concerned about harmful magic than they were about the involvement of demons or the Devil. We know this because when people accused others of witchcraft, they almost invariably focused on claims that the accused had used magic to injure others rather than on claims of demonic or diabolical collaboration. This likely reflected contemporary beliefs and fears about folk magic as well as more pragmatic concerns about direct harm or injury.

Witch hunts almost always began with an individual being accused of *maleficia* by his or her neighbors, usually triggered by specific events or misfortunes suffered by a community. This might include a hailstorm that destroyed local crops, mysterious illnesses that sickened people or livestock, or the death of children. Once the accusation was made, usually to a local official such as a mayor, landlord, or magistrate, it brought the suspected witch to the attention of higher judicial authorities. These might be judges appointed by a secular government or ecclesiastical officials such as inquisitors, but whether their authority derived from secular or religious institutions, these judicial figures were usually educated in and familiar with witchcraft theory. Often, ordinary people were exposed to that theory in the course of investigations or trials carried out by the authorities. For example, they might read or hear the specific charges laid against a neighbor accused of witchcraft and thereby gain some understanding of the practices in which witches were supposed to engage, such as consorting with demons or flying to the sabbat.

For the accused, it is likely that they would first encounter the theories held by the educated classes while they were being questioned. We know this because detailed records of judicial processes became more and more common from the late Middle Ages. In studying those records that have survived, historians have found that inquisitors and other judicial authorities sometimes asked leading questions – "Isn't it true that you flew to the sabbat and there met with the Devil?" – that helped the accused understand what the inquisitor wanted to hear.

The interrogation or questioning of those accused of witchcraft did not merely introduce them to the beliefs of witchcraft theorists, however; the use of torture often meant that the accused also corroborated or confirmed those beliefs. Judicial torture gradually started to appear in jurisdictions across Europe in the fourteenth and fifteenth centuries as communities adopted what historians call the *inquisitorial proced-ure*, a system of legal processes and practices that gave greater author-ity and powers to judicial officials in the instigation and prosecution of crimes. Many officials believed that torture was necessary in the pur-suit of suspected witches because heresy and witchcraft were both concealed or hidden crimes. Direct or eyewitness evidence was not always possible to obtain, so establishing innocence or guilt rested solely on the testimony of the accused and torture was seen as a reliable means of acquiring that testimony.

The use of torture in pursuit of a confession was tightly regulated at first, with numerous monarchs and governments passing laws restrict-ing how and when torture could be employed, but in parts of Europe where centralized or governmental authority was weak and the judicial systems in smaller territories and city-states operated with little over-sight, authorities often used torture freely, without regulation or restraint. This was particularly true in the Holy Roman Empire, Switzerland, and parts of eastern France, all of which were composed of territories and communities that resisted the imposition of centralized authority. As a result, these parts of Europe witnessed some of the largest and most intense witch hunts. By contrast, the Spanish and Roman Inquisitions were far more cautious when it came to the use of judicial torture; under their jurisdiction, torture was applied more rarely and was tightly regulated when it was used at all. As a result, we find significantly fewer cases of suspected witches being executed in Spain and Italy at the height of the hunts in the sixteenth and seventeenth centuries.

The role of torture in reinforcing witchcraft theory becomes obvious when we find that particular ideas, such as the role of the Devil in the pacts supposedly made by witches, usually appeared in the testimony of individuals only *after* they had been subjected to rigorous questioning or torture. At the same time, in parts of Europe where judicial torture was less common, such as England, Spain, and Italy, we find significantly fewer people confessing to flight, Devil worship, or other elements of learned witchcraft theory. Instead, people in those countries were much more likely to be accused of *maleficia* alone, without overt reference to demonic or diabolical forces, and as a consequence their punishment was often less severe than those accused of Devil worship.

Brian Levack and other historians of witchcraft have pointed to a clear correlation between the increasing reliance on judicial torture and the scale of the witch hunts: as torture became more widespread, more people confessed to being witches. The practice of arresting and torturing those implicated by suspected witches, usually without any other corroborating evidence, almost invariably led to what Levack calls "chain-reaction hunts" in which dozens or hundreds of people both confessed to being witches and accused others of the same. Most of these hunts took place in what is now Germany, Austria, and Switzerland between the late sixteenth and mid-seventeenth centuries, and they led to the imprisonment and execution of women, men, and children as more and more people desperately implicated their neighbors and even their own families as fellow witches.

This pattern makes a grim kind of sense. Where judicial torture was largely unregulated, it tended to result in false confessions as those accused of witchcraft – perhaps prompted by questions from their interrogators – said whatever was necessary to make the torture stop. In turn, this meant that inquisitors and magistrates heard what they expected to hear, which only made them more determined to find similar kinds of evidence in the next person they questioned. In this way witchcraft theory became essentially self-fulfilling, at least within the confines of the judicial questioning of suspected witches. As a result, there was no real impetus to challenge a theory that was continually corroborated by the accused. This may be why ideas about witchcraft were not widely or successfully challenged until the late seventeenth century, when most governments either heavily restricted or abandoned the use of judicial torture. This also explains why there were places in Europe where relatively few people were found guilty of witchcraft and executed for their crimes. Compared to central Europe, for example, England witnessed very few executions of witches, due in large part to its unique judicial system in which torture was almost never employed and verdicts were handed down by juries rather than by a single judge or magistrate.

The continuous corroboration of witchcraft theory through torture is important because the theorists needed to convince or persuade different audiences to believe their claims. Their ability to instigate and sustain full-fledged witch hunts required the support of the ruling classes, who would cooperate only if they believed that witchcraft posed a serious threat to the stability of their region. Rulers needed to be persuaded that witches were doing more than harming people with magic; they also had to see witches as fundamentally opposed to the very foundations of the Christian state and as a serious threat to order and morality. For the common people, diabolism (meaning the direct involvement of the Devil) was not a primary concern at the start of the witch hunts. They tended to be more worried about the harmful effects of magic practiced by a malicious witch rather than questions of how or why that magic was possible. Nonetheless, both the ruling classes and the common people came to accept the claims made by witchcraft theorists – that witches, under the influence of the Devil, were working to undermine and overturn Christian society – and eventually integrated those claims into their own understandings of witchcraft.

The End of the Witch Hunts

As far as historians know, the last judicial execution of a witch took place in Switzerland in 1782. Long before this, however, the witch hunts were already in decline across Europe. Some regions, such as

England and the Low Countries (the modern-day Netherlands), had ceased holding trials of suspected witches more than a hundred years earlier, and in Portugal the last judicial execution of a witch occurred in 1626, more than 150 years before the last execution in Switzerland. Just as the witch hunts varied from place to place, so did their decline. Generally speaking, witchcraft trials ended anywhere between the late seventeenth century and the last decades of the eighteenth century.

An important factor for this decline is that the appetite for judicial torture decreased sharply from the middle of the seventeenth century. In fact, across much of Europe judicial torture was either strongly discouraged or prohibited altogether between the mid-seventeenth century and the end of the eighteenth century, reflecting changing social and judicial attitudes toward the practice. Perhaps it is not surprising that, when suspected witches no longer faced the threat of torture, they became far less willing to corroborate the accusations of inquisitors or accuse their friends and neighbors of being witches. This fact alone demonstrates the troubling degree to which torture and other threats of violence encouraged and supported the witch hunts for hundreds of years.

Social and economic conditions also improved in Europe from the middle of the seventeenth century. Food prices gradually fell, periods of famine became more rare, the threat of armed conflicts diminished, and rates of infant mortality decreased. The upheavals that had characterized the period between the fourteenth and sixteenth centuries gave way to a Europe that was generally more stable and prosperous than before. Meanwhile, the Catholic Church had been transformed first by the Protestant Reformation and then by the period known as the Catholic Reform (sometimes called the Counter-Reformation). At the heart of this new reformation was the Council of Trent, which took place between 1545 and 1563 and which enacted a series of important changes and reforms to Catholic doctrine and practice. The wars of religion which had raged across much of Europe in the sixteenth and seventeenth centuries were largely concluded by the Peace of Westphalia in 1648, reducing religious tensions across Europe as well as eliminating widespread conflict. Anxieties about widespread heresy were no longer as pronounced in the sixteenth and seventeenth centuries as they had been in the fourteenth and fifteenth centuries, and both Catholic and Protestant theologians began to question openly whether demonic or diabolical influence was really as widespread as was once believed. Ideas about the role and presence of God in the universe also began to change in the seventeenth and eighteenth centuries, a shift that might have made theologians less interested in finding tangible proof for God in the sexual relations between witches and demons.

Though the formal or official persecution of accused witches declined into the eighteenth century, those suspected of witchcraft were still subjected to informal "justice" in the form of lynchings, exile, or other violent acts committed by members of their communities. Such things already had taken place for hundreds of years, but as formal trials disappeared lynchings and other violent acts of retribution became either more commonplace or more noticeable. Such practices have not disappeared in the modern world, unfortunately; there are still cases of violence and persecution against those suspected of witchcraft in parts of Europe, Africa, and southeast Asia. Nor has the idea of the witch hunt faded away entirely. The rise of McCarthyism in the United States in the 1950s, in which Senator Joseph McCarthy (1908–57) organized the House on Un-American Activities Committee to hunt down communists hiding on American soil, led the playwright Arthur Miller to write his now-famous play *The Crucible* in 1953, which highlights the parallels between McCarthy's activities and the witch trials in Salem, Massachusetts in the 1690s.

In the modern West, witchcraft remains a part of popular culture. Witches, both good and evil, appear in TV shows, movies, and video games. New Age spirituality, particularly the tradition of Wicca, has reclaimed the word "witch" as a positive title, and modern-day witches have started to demand public recognition and respect. On the flip side, anxious Christians have found traces of witchcraft in pieces of popular culture like the *Harry Potter* novels, leading them to label these pieces of popular culture as Satanic and diabolical much like inquisitors did centuries ago. While the European witch hunts ended centuries ago, they still cast a long shadow today.

3 | *Magic, Medicine, and the Microcosm*

In this chapter we explore how occult philosophies and practices were applied to the human body. Both magic and religion have played important roles in medicine for many thousands of years, but in the sixteenth and seventeenth centuries increasing numbers of educated Europeans turned to magic, alchemy, and astrology in their attempts to heal and strengthen the human body. Illness, injury, and disease afflicted everyone, regardless of wealth or class, and we have already seen that traditions of folk magic had been used to treat illness and injury for hundreds, if not thousands, of years. The witch hunts may have cast some of those practices in a negative or potentially dangerous light, but they persisted in rural as well as urban communities even as philosophers and physicians debated new and exciting remedies inspired by the hidden forces and powers of the natural world.

This chapter begins with an overview of medical theory and practice, including the state of medical education in premodern Europe and attempts made by some humanist reformers to change how medicine was taught. We then look at an old idea known as the *microcosm* or "tiny world," which refers to the belief that the human body is a reflection or miniature copy of the larger universe. The connection between these two worlds is subtle and sometimes difficult to explain, but it means that the movements and phenomena that exist in the larger universe have an effect on the tiny world that is the human body. This idea has strong links to the theory and practice of astrology, which attempts to explain and predict how the heavens affect life here on Earth. Together, the idea of the microcosm and the practice of astrology were central to how many educated people thought about medicine and its ability to heal the body.

This chapter also introduces the Swiss medical reformer known as Paracelsus (1493–1541). He was a controversial figure, frequently mocked and reviled during his lifetime but hailed as a pioneer of medicine after his death. Paracelsus believed that nature itself was the key to treating illness or injury, and that the task of the physician was to investigate the natural world and uncover its many secrets. If we keep in mind that magic was seen primarily as a kind of applied science – an attempt to unlock the mysteries of nature and then apply them in specific and useful ways – then the efforts of Paracelsus do look very much like a kind of magical medicine. While he may not have used the word "magic" to describe his ideas, the beliefs and philosophies of Paracelsus will help us understand more clearly how magic, religion, and medicine overlapped in this period, each influencing the others in important ways.

Finally, this chapter examines one of the most fascinating remedies from early modern Europe, a cure that was attributed to Paracelsus but which really had very little to do with him. It was known first as the weapon salve and, later, as the powder of sympathy. These two remedies were thought to heal wounds quickly, painlessly, and without infection over distances of many miles when applied only to traces of the wounded patient's blood – if you applied the salve to a weapon that had injured someone, for example, it could heal that person even if they were far away. These cures demonstrate how blurred the line was between magic, witchcraft, and medicine. Scandals ensued and reputations were ruined when some people fell on the wrong side of that line. Their stories will help us understand how early modern people tried to distinguish legitimate ideas and practices from those considered illegitimate. Sometimes, as in the case of these particular remedies, it was difficult to tell the two apart.

Learned Medicine in Premodern Europe

Premodern academic medicine was founded primarily on the theories of two ancient writers: Hippocrates (c. 460–c. 370 BCE), a Greek physician, and Galen (129–c. 210), another Greek who lived in the Roman Empire some 500 years after Hippocrates. Together, Hippocrates and Galen provided most of the theoretical structure for European medicine for close to 2,000 years, though their ideas were supplemented and improved upon by Islamic medical writers such as Ibn Sīnā or Avicenna, a Persian physician who died in 1037 CE and who may have become a greater influence on later generations than either Hippocrates or Galen. It is interesting that these ancient roots of Western medicine have persisted in some forms today. Physicians in many countries still swear to uphold a version of the Hippocratic Oath, which regulates the behavior of medical practitioners and which claims to date back to Greek antiquity.

Hippocrates and his followers taught that disease was natural, not the work of vengeful gods or other supernatural phenomena, and arose from imbalance. The healthy body existed in a state of equilibrium, with opposing elements balanced in a particular way unique to that individual. Hippocratic writers believed that these opposing elements were *humors*, substances within the body that had different characteristics. There were four humors – blood, phlegm, black bile, and yellow bile – and they were understood to differ in terms of both temperature and moisture. Blood was hot and wet, while black bile was cold and dry; phlegm was cold and wet, while yellow bile was hot and dry. Disease occurred when the balance of these humors was disrupted, and the task of the physician was to bring them back into equilibrium.

Humor	Characteristics	Element	Organ	Temperament
Blood	Hot and wet	Fire	Liver	Sanguine
Phlegm	Cold and wet	Water	Brain	Phlegmatic
Yellow bile	Hot and dry	Air	Spleen	Choleric
Black bile	Cold and dry	Earth	Gallbladder	Melancholic

This is known today as the *humoral system* of medicine, and it was the predominant medical theory in Europe for close to 2,000 years. Each humor was connected to a particular organ within the body where it was thought to originate or where it was most abundant. The humoral system also described different *temperaments* (behaviors, emotions, even physical appearances) that were the result of an overabundance of a particular humor. The longevity of this system can probably be explained in part by the fact that it fit neatly within and around other, existing systems; for example, the humors were usually connected to the four classical elements of earth, water, air, and fire, which themselves were part of many other ancient philosophies. The idea that health is a state of equilibrium and that disease is the result of disruption or imbalance also makes a kind of intuitive sense. Injury and illness certainly feel very disruptive to those who suffer from them.

In order to restore balance, the physician prescribed a regimen of treatment that might include medicines or other substances but which also regulated diet, exercise, and sleep. Someone diagnosed as choleric, for example, would suffer from an overabundance of yellow bile and exhibit characteristics that were hot and dry, and so the physician would prescribe certain foods that were thought to be "cold" and "wet" in their internal properties as well as forms of exercise that, in this case, might include swimming in cool water and avoiding the hot sun. By altering both the inside of the body as well as its outside environment, the hope was that the humors would fall back into balance and restore the patient to health.

The humoral system of medicine persisted well past the Renaissance, into the sixteenth and seventeenth centuries, though medical reformers like Paracelsus challenged and critiqued this system in different ways. One reason for the longevity of these medical theories is that they supported practices that actually healed the sick or, at the very least, did not kill them outright. For example, bloodletting or *phlebotomy* was a common medical practice for centuries; medical practitioners would tell their patients to relieve themselves of a quantity of blood in order to cure literally dozens of different ailments. Humoral theory suggested that an excess of blood might lead to internal imbalance and that purging some blood would help, and while some patients were undoubtedly subjected to bloodletting when they were too weak to endure it, for most people it caused no lasting damage. Here was a medical practice that satisfied the prevailing theory and, at the very least, did no obvious harm. No wonder it was still commonly prescribed into the eighteenth century.

But who exactly were the medical practitioners offering this advice? We have already seen that European communities often turned to individuals like midwives, cunning-folk, and other practitioners of folk medicine to seek relief from certain ailments or advice about particular problems. For more serious injuries and diseases, however, people generally consulted three main groups: physicians, barber-surgeons, and apothecaries. Physicians were usually university-educated, and were licensed by local guilds or associations to practice medicine. They stood at the top of the medical hierarchy and normally dealt with serious illnesses or with internal diseases and injuries. Barber-surgeons, by contrast, dealt almost exclusively with injuries and other problems on the outside of the body or that could be addressed fairly easily: tooth-pulling, broken bones, cataracts, as well as cutting hair (Figure 3.1). While physicians were literate and licensed, the typical

Figure 3.1 Sixteenth-century engraving of a barber-surgeon removing a tooth.
Photo by Culture Club/Getty Images

barber-surgeon was a craftsman who learned his trade as an apprentice to a more experienced person before striking off on his own. According to many physicians, barber-surgeons were an inferior but necessary form of practitioner, dealing with routine or messy problems that the average physician did not wish to address. Finally, there were the apothecaries, who were not medical practitioners themselves but who provided the hundreds of ingredients and substances required to make medical remedies. They were the premodern equivalent of our modern-day pharmacy, and they played an important role in the medical economy of the day.

For the average European, consulting a physician was an expensive proposition, which means it rarely happened. Those who could afford it were usually instructed to present a flask of their urine; that and the pulse were the two main diagnostic tools used by the physician to determine what was happening inside the body. Based on these indicators, the physician would prescribe a remedy or regimen designed to return the body to health. Uroscopy, or the examination of urine, was considered so reliable by some physicians that they diagnosed their patients without actually seeing them; a sample of urine, brought by

Figure 3.2 A physician examining a urine flask brought by a servant, from an 1849 reproduction of a fifteenth-century engraving.
Photo by Culture Club/Getty Images

messenger, was considered sufficient to determine illness and prescribe a remedy (Figure 3.2). Though popular, this practice was criticized sharply by some who saw it (perhaps rightly) as a way for the physician to make some easy money without leaving the comfort of his own home.

Equally problematic, at least from our perspective, was the way in which many physicians were trained. Their education usually consisted of university lectures in which a professor simply read aloud from a respected medical text – for example, the *Canon* of Avicenna – and students quietly copied down the text for their own use later. It was entirely possible for a student to spend several years learning medicine without ever once witnessing the dissection of a human body, an idea that would be unthinkable in the context of modern medical education. The problem was that human cadavers were not widely available in the early modern period. Occasionally a prisoner would be executed and their body made available for dissection; the great humanist anatomist Andreas Vesalius (1514–64) once wrote of watching a man have his head chopped off and then running up to grab it from the basket into which it had fallen, before hurrying home to dissect it while it was still warm. In those rare cases when a body was made available to medical students, the dissection would last for several days as a barber-surgeon carefully opened the body's many layers, though it was not uncommon for the physician-professor to be reading a passage about, say, the skeleton while below him, on the floor of the lecture theater, the barber-surgeon displayed the liver or the spleen.

This uncoordinated procedure was described and sharply criticized by Vesalius in the work that would define his career and cement his place in history: *De humani corporis fabrica* or "On the fabric of the human body," known by most scholars today simply as the *Fabrica*. This was an enormous atlas of the human body that first appeared in 1543, and it was the most comprehensive study of human anatomy produced up to that point. Vesalius was unhappy with the division of labor that characterized medical education – the physician-professor lecturing while the barber-surgeon cut open bodies – and he cast himself as a humanist reformer of medicine, inspired by the works of Galen but committed to correcting their many mistakes. Galen, who had lived and worked in the ancient Roman empire, had never opened a human body because dissection was culturally forbidden at that time; instead, he had dissected dogs and apes and then applied his findings to the human body. Vesalius, by contrast, presented himself as the first of a new breed of medical expert. He was highly learned but also skilled with the knife, capable of lecturing and dissecting simultaneously. The image on the front of the *Fabrica* shows an authoritative Vesalius lecturing next to an open body while around him spectators push closer to get a good look. On either side of the dissecting table, at the very front of the image, are two towering figures dressed in classical togas and sandals; these are meant to be Aristotle (whose works on zoology were well known in Vesalius's time) and Galen himself, both watching attentively as Vesalius schools them in anatomy. Behind Galen is a small monkey, a reference to his experience dissecting apes (Figure 3.3).

The *Fabrica* marked the beginning of what we would understand as a rigorous and empirical approach to the study of anatomy. Vesalius probably hired artists from the workshop of the great Renaissance artist Titian (c. 1490–1576) to provide the illustrations, and the *Fabrica* is filled with hundreds of beautiful images that were unlike anything seen before in an anatomical text. Vesalius discussed every known part of the body and refuted many errors that had descended from Galen, though he still clung to some ancient misconceptions

Figure 3.3 Title page from *De humani corporis fabrica* by Andreas Vesalius, 1543.

himself, including those concerning the structure of the liver and the circulation of the blood. While the influence of the *Fabrica* is important, however, so too was Vesalius's criticism of medical education. His ideas circulated widely after the *Fabrica* appeared in print, and his explicit challenge to Galen's ancient authority was a crucial step in the more widespread reform of medical theory and practice that took place over the next 150 years. Like his contemporary Nicolaus Copernicus, who we will meet in the next chapter, Vesalius made it permissible to question the legacy of antiquity and, in some cases, to overturn ideas that had persisted for many hundreds of years. Also like Copernicus, Vesalius found it impossible to divorce his beliefs entirely from the ancient ideas he had been taught. Both men were poised somewhere between the ancient and the modern, unable to reject the one entirely but helping to build the other.

The Microcosm, the Macrocosm, and Astrology

The idea of a relationship between the microcosm and the macrocosm stretches back to antiquity and is not just a European phenomenon; similar ideas have existed in many cultures for a very long time. At its most basic, it assumes that there are two "worlds" that mirror each other. The first (and larger) is the macrocosm, which encompasses the whole of the universe or, in some interpretations, the Earth; the microcosm, then, is a much smaller reflection of that larger world. This works in the other direction, too: the wider universe also reflects the smaller microcosm. The two are linked in any number of arcane and mysterious ways, but it is the fact of their connection that matters.

The microcosm could be anything that mirrors the structure or qualities of the universe, but for many early modern people the true microcosm was the human body. Physicians, astrologers, philosophers, and alchemists all taught and believed that the body reflects and is affected by the wider universe (Figure 3.4). The connections between the two were usually metaphorical – for example, comparing the heart to the Sun and the circular motion of the planets to the circulation of the blood through the body – but they expressed an idea

Figure 3.4 Man as microcosm, surrounded by the macrocosm. From Robert
Fludd, *Utriusque cosmi...historia*, c. 1617.
Photo by Oxford Science Archive/Print Collector/Getty Images

that was central to esoteric or magical philosophies like hermeticism:
everything is connected, or, in the aphorism quoted by innumerable
hermeticists, "as above, so below." Indeed, many different systems
and philosophies have as a central or defining idea that different
objects or substances are connected with one another in hidden and
mysterious ways, and it makes sense to understand the macrocosm/

microcosm theory as falling into this same category. There might even be a spiritual or religious connotation to the idea that the body is connected to the wider universe and reflects that universe in some respects.

These connections between the body and the universe made particular sense to premodern Europeans because they, like people in many other cultures, embraced a belief in astrology. While astronomy is the scientific or mathematical study of the heavens, which means everything beyond the Earth, astrology uses the movements and positions of planets and other celestial objects to predict and understand events here on Earth. Some people today refer to astrology as a science, but it is perhaps more accurate to understand astrology as *systematic* rather than scientific. It has sets of rules and beliefs that date back centuries, in some cases, but these systems of rules are not scientific in the way we understand the word today.

The central belief or idea at the core of astrology is straightforward: What happens in the celestial realm affects and influences the terrestrial realm. If we believe that we live in a geocentric universe in which everything literally revolves around the Earth, it makes an intuitive kind of sense that those celestial bodies up there would influence people and events down here. In fact, the idea that the heavens have an influence on us and our lives is so compelling that it is an idea shared across many different cultures and times. Many millions of people alive today believe in astrology, either because it is an important part of their culture or because it seems to provide answers and explanations about their lives. It is an interesting fact that even our modern advances in astronomy have had relatively little impact on astrological beliefs.

Perhaps the best-known aspect of astrology is the horoscope, which explains how astrological influences directly affect the life of a particular person. You might know what your "sign" is, based on when you were born (for example, I'm a Sagittarius). These refer to the twelve (or, according to some, thirteen) signs of the Western zodiac, which is the band of constellations through which the Sun appears to move over the course of a year. Your astrological sign represents the constellation in which the Sun rose on the day you were born. When many of us in the Western world read our horoscopes, they are based on this system of zodiac signs, though other cultures have different associations with specific celestial objects, constellations, and abstract figures.

The preparation of horoscopes was an important endeavor in early modern Europe. If you were wealthy enough, you could hire an astrologer to prepare your horoscope or, as it was known at the time, your *geniture* (derived from the Latin word for "birth"). This geniture would provide you with information about your history and your future life, including your family, your fortunes, and your health, and preparing it was a difficult and highly technical task. Early modern genitures were typically divided into twelve houses, each governing a different aspect of someone's life, and the astrologer usually consulted complicated tables published as part of astronomical almanacs in order to calculate the influence of specific planets on these different houses. This was not a simple task, and it certainly was not like the horoscopes we find today in which a prediction might apply to millions of people who share the same astrological sign. Genitures produced by astrologers in premodern Europe were unique to the individual, and creating them required both training and expertise.

If we accept, as premodern people did, that the body and the wider universe were connected, then we can understand why astrology was such an important part of premodern medicine. Someone who went to university and studied for a medical degree almost certainly would have been trained in astrology, something which made perfect sense to people living in early modern Europe. If the body is a reflection of the cosmos, then if we want to heal injuries or disease we need to understand both how the cosmos works and how it affects people and objects here on Earth. Medical practitioners associated signs of the zodiac with different parts of the body – Sagittarius, for example, ruled over the thighs or legs, while Gemini was associated with the arms or lungs and Leo with the heart or the back – and sometimes used these connections to dictate treatment (Figure 3.5). There were cases of surgeons refusing to operate on a specific part of the body unless the heavens were aligned with the corresponding zodiac sign, and it was not uncommon for learned physicians to cast their patient's horoscope as part of their diagnosis. This was simply one way in which medical practitioners tried to achieve the best outcomes for their patients.

There were also strong connections between the humoral theory of medicine and astrological beliefs about the planets. Both were based on the idea of opposing qualities advocated by Hippocrates and many other ancient authorities, according to which all substances could be understood in terms of heat (hot or cold) and moisture (wet or dry).

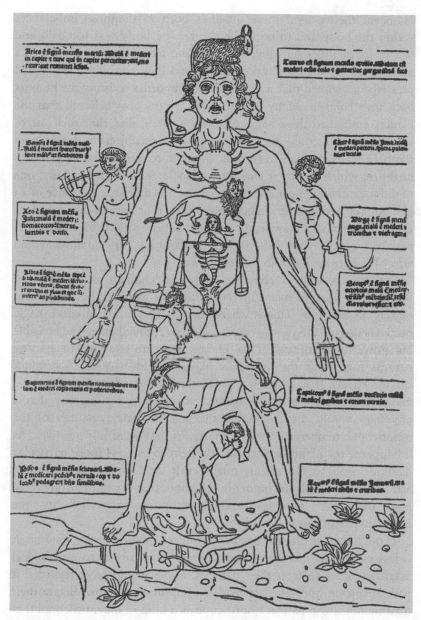

Figure 3.5 Zodiac signs and their astrological associations with different parts of the human body. From Joannes Ketham, *Fasciculus Medicinae*, 1495.

Photo by Bettmann/Getty Images

Humoral theory used these qualities both to diagnose imbalances within the body and to prescribe a cure. If a body was too cold and wet in its qualities (creating a phlegmatic temperament as well as a number of associated physical afflictions) then the physician would prescribe a regimen that was hot and dry in order to bring the humors back into balance. Many educated Europeans also believed that the planets embodied these same qualities as well, and that their astrological influence could impart those qualities to objects and people on Earth. For example, Saturn was thought to possess cold and dry qualities, which is why it was often associated with melancholy, one of the four main temperaments. If Saturn had a prominent place in an individual's geniture, it might explain why that person had a melancholic disposition. Likewise, if the planet Saturn was visible in a certain part of the sky, many believed that its cold and dry qualities would affect life here on Earth in different ways.

The planets were a constant source of change; as they moved through the heavens their different influences were felt by people and objects all the time. This is where the practice of astrological magic becomes useful, and perhaps particularly so for medicine. The invisible but powerful influences of the planets were an example of another natural force that the ingenious magus could harness to produce specific effects. If someone was particularly afflicted by melancholy, a magus might seek to create balance in much the same way that a physician would, by countering the cold and dry qualities of the melancholic temperament with influences that were hot and wet. As it happens, the planet Jupiter was thought to possess those particular qualities, and so the magus might craft an amulet inscribed with astrological symbols for Jupiter in the hopes that it would attract these hot and wet qualities and thereby bring relief to the melancholic sufferer.

Though the use of astrology in premodern medicine was common, it is less clear how often physicians would have turned to astrological magic in order to treat their patients. Some would have regarded it with suspicion and relied instead on genitures alone to dictate their treatment, using a patient's horoscope as a kind of diagnostic tool that provided useful information about that person's temperament and other influences on their health. Astrological magic was a different thing altogether, requiring the practitioner to harness the unseen forces and emanations of the planets to heal their patient rather than relying solely on a standard regimen of care. While some reputable practitioners did use magic in their treatments, a great many unlicensed and unregulated practitioners, usually called quacks or mountebanks, tricked the gullible and the desperate into purchasing talismans and trinkets supposedly imbued with magic. This association with fraud is one reason why magic had such a problematic reputation in this period. At the same time, however, the huge market for such cures demonstrates that it had a powerful hold on the collective imagination of premodern Europe throughout this entire period. Even if some viewed magic with distrust and suspicion, many others saw it as a useful way to treat illness and ward off disease.

This belief in the power of both astrological and magical medicine was not confined to rural backwaters and the realm of "peasant superstition," as some of the educated elite liked to claim; it pervaded European society at every level. For example, the extensive medical casebooks of the Elizabethan practitioner Simon Forman (1552–1611) and his protégé Richard Napier (1559–1634) reveal how they used astrology to diagnose and cure many wealthy and well-connected patients around the turn of the seventeenth century. Neither man was trained as a physician Forman never completed his university studies and Napier was a clergyman. Yet, both established themselves as healers of some repute. Forman's writings in particular demonstrate an avid interest not just in astrology but in alchemy and magic as well. His willingness to employ such practices earned him the contempt of more conventional practitioners in the Royal College of Physicians as well as a reputation in some circles as a quack and a fraud. Nonetheless, the astrological foundations of the medicine practiced by Forman and Napier would have been commonplace and generally accepted in this period.[1]

The microcosm/macrocosm system encouraged individuals to see their different worlds as closely intertwined, linked by mysterious and invisible correspondences. One might argue that this way of thinking also encouraged a more holistic vision of the individual – people saw themselves as connected to the world around them and understood their own bodies and their health as intimately tied to the larger universe.

---

[1] To learn more about Forman, Napier, and their work, see Lauren Kassell, Michael Hawkins, Robert Ralley, John Young, Joanne Edge, Janet Yvonne Martin-Portugues, and Natalie Kaoukji (eds.), "Casebooks," *The casebooks of Simon Forman and Richard Napier, 1596–1634: a digital edition*, https://casebooks.lib.cam.ac.uk.

Figure 3.6 A sixteenth-century portrait of Theophrastus Bombastus von Hohenheim, or Paracelsus.
Photo by ullstein bild/ullstein bild via Getty Images

Astrology gave Europeans a sense of their place in the cosmos, but at the same time it empowered them with the belief that they could predict the course of their lives and make sense of its inevitable ups and downs. More particularly, the practice of astrological magic offered the hope that the individual could harness the hidden powers of the universe to protect their health and improve their life.

Paracelsus

In 1541, Andreas Vesalius was hard at work on the *Fabrica* and dreaming of a secular reformation of medicine that would be as sweeping and widespread as the Protestant Reformation still gripping Europe. In that same year, another would-be reformer of medicine died in poverty, his life's work largely overlooked by his contemporaries. It would be another forty years before his name became widely known, when it was attached to a new theory of medical practice that shook the foundations of medicine as it was then known. This was Paracelsus, whose ideas inspired the movement known as *Paracelsianism* and who remains a controversial and fascinating figure more than 450 years after his death (Figure 3.6).

His given name is usually written as Theophrastus von Hohenheim or, in some cases, Philippus von Hohenheim, though admirers expanded it until it read more like a title: Philippus Theophrastus Aurelius Bombastus von Hohenheim. The name by which he is best known, Paracelsus, is something of a mystery, but historians believe that it was inspired by the classical Roman medical writer Celsus (c. 25 BCE–c. 50 CE). The prefix "para-" that he added to that ancient name has multiple meanings in Latin, including "beyond," leading some to speculate that this was a not-so-modest attempt to claim a knowledge of medicine that was greater than that of Celsus. Others believe that it was more of an homage. Whatever the case, Paracelsus was not shy about touting his own abilities as a healer, and he had little respect for much of the medical knowledge that had been passed down from antiquity.

Many implausible, even extravagant claims have been made about Paracelsus, such as that he was the inventor of chemotherapy, toxicology, and homeopathy. As with quite a few stories about Paracelsus, all of these "facts" are sort of true and yet sort of *not* true. It is true that he advocated for a type of medicine that relied on metals and minerals to a significant degree, and taught that chemistry was the art by which these substances might be made useful to the physician. This way of curing was quite different from the botanical or plant-based remedies that had long been the core of Western medicine, but was not "chemotherapy" except in the loosest sense of the term. Likewise, his observations of toxic substances and their effects on the body were not "toxicology," which implies a level of systematic study that Paracelsus rarely achieved.

He also taught that "like cures like," which happens to be the main idea behind modern homeopathy in which extremely diluted amounts of certain substances are administered to patients in an effort to mimic the symptoms of whatever is ailing them. Paracelsus advocated for his doctrine of "like cures like," however, in opposition to the idea in traditional medicine that the diseased body was out of balance and so required the application of opposites to return it to health – if it was feverish then medicines made from "cold" ingredients were prescribed, and so on. Paracelsus believed that a fever required "hot" medicines rather than "cold" ones, which is not the same thing as homeopathy as we understand it today.

Unfortunately, we know only a limited amount about Paracelsus's life. In part this is because he was not very forthcoming about his origins, and in part because during his lifetime he gained a reputation for exaggeration and outright fraud, so at least some of his claims were regarded with suspicion by contemporaries. After his death admirers and followers fabricated portions of his biography and even created false letters and other documents, meaning that historians today are not always sure what to believe. We do know a few basic facts, however. He was born in what is now Switzerland, and his father was also a physician who ensured that his son received a broad education in medicine, natural philosophy, and the liberal arts. He is thought to have received his medical degree around the year 1516, though even this fact is disputed by those who claim that Paracelsus never received a medical degree at all. For the rest of his life he wandered across central Europe and perhaps much further afield as well; there are hints that he spent time in what is now Turkey and perhaps ventured west to Britain as well, though there is limited evidence to support this. What we can say with certainty is that Paracelsus never stayed in one place for long. Sometimes this was because he irritated or outraged a great number of important people in towns and cities all over Europe; more than once he skipped town just before he would have been violently evicted. But he also seems to have been a born wanderer who was eager to see new places and have new experiences.

Paracelsus did gain some repute as a practitioner of medicine during his lifetime. He cured important people who recommended him to their friends and acquaintances, and even taught medicine for a short time in Basel, though that particular episode ended disastrously. Paracelsus, however, had very different ideas about the practice of medicine than most of his contemporaries. Paracelsus rejected the standard view that disease could be cured by the administration of opposite qualities, and while traditional medicine relied on the Galenic philosophy of the four humors and their links with the four elements, he taught that everything in the world was composed of what he called the *tria prima*, which we might translate as the "top three." These were principles even more fundamental than the four classical elements, and according to Paracelsus they were Salt, Sulphur, and Mercury. These were not the tangible, visible substances that we might associate with those words, however; Paracelsus was not talking about table salt or liquid mercury. Instead, he understood the *tria prima* as intangible essences present within all substances to varying degrees.

Each of the three principles supplied different characteristics to a particular object; for example, he equated Salt with solidity and matter, providing the tangible qualities of an object, while Mercury provided life or spirit to living things. Paracelsus often illustrated the *tria prima* with the example of a burning piece of wood – as the wood burned it produced each of the three principles: the ash left behind was the Salt, the burning fire itself was Sulphur (the most volatile principle), and the smoke given off was Mercury. In fact, fire occupied a central place in Paracelsian theory and practice. Paracelsus envisioned it as the best, perhaps only means of breaking down different substances into their component parts, which the skilled physician could then use or recombine to create powerful cures and remedies.

Another way in which Paracelsus differed from other physicians was in his ideas about medical education and practice. He insisted that good medicine sprang from the innumerable secret and beneficial properties that existed in nature, and that the proper task of the physician was to seek out these properties and harness them to heal the body. This is why Paracelsus had such a low opinion of traditional medical education, in which aspiring physicians sat in lecture halls and listened while a professor read aloud from a famous medical text like the *Canon* of Avicenna (a copy of which Paracelsus supposedly, and scandalously, threw into a bonfire to demonstrate his contempt for conventional medicine – this was the disaster that ended his short teaching career at Basel). He argued vigorously that reading books could never educate someone about the hidden properties of nature; instead, the physician should wander the world as Paracelsus himself did, studying plants and minerals and testing their properties with his own hands. Only by uncovering those secret properties and putting them to use could the physician truly heal.

Of course, those secret and beneficial properties did not arise spontaneously. Paracelsus was a deeply pious man who believed that God had implanted these properties in all natural substances. After all, God is benevolent and would not require people to suffer from illness or disease without also providing the means to cure them. For Paracelsus, the whole of nature was like one enormous pharmacy or apothecary's shop, filled with medicines just waiting to be uncovered by the pious and dedicated physician. This begs an obvious question, however. If God hid these medicines in nature, how are we supposed to find them? Again, God's benevolence would ensure that these medicinal properties *can* be found – He would not make it impossible for us to find and use them – and according to Paracelsus, nature itself was designed in such a way that it reveals its secrets to the dedicated searcher.

The underlying idea that drove this thinking, and that was embraced wholeheartedly by others after Paracelsus died, is now called the *doctrine of signatures*. Paracelsus believed that God had left clues in nature, signatures that hinted at the hidden properties within plants, minerals, and other substances. For example, according to this doctrine a plant with heart-shaped leaves probably possesses some quality or property useful in treating ailments of the heart. Walnuts, when extracted from their shells, look like brains, which is a hint that they possess some property that is beneficial in treating ailments of the brain or head. According to Paracelsians, nature is full of these hints and clues, scattered there by God and just waiting for the industrious and curious physician to find them. This, again, is why Paracelsus had such disdain for those who learned medicine by reading books. Remedies for disease did not exist in books; they could be found only in nature, and discovered only by the physician who used his eyes and hands to investigate the world.

This quest for nature's hidden secrets is very much in the same vein as our previous discussions of magic, though it is unlikely that Paracelsus would have embraced our use of that term. For him, medicine was not a magical practice but one that straddled the line between natural philosophy and theology. Nevertheless, the practices he advocated were virtually identical to those employed by practitioners of magic, hermeticism, and alchemy in this period, since they too were interested in finding and using nature's hidden secrets. In fact, Paracelsus was a vocal proponent of alchemy, which he viewed as the art of uncovering nature's secrets through the application of fire, and he argued that it was the best way to reveal and utilize the hidden virtues and qualities of medicinal substances. After his death many of his followers found innovative ways to combine his medical theories with chemical and, more specifically, alchemical practices.

Paracelsus's legacy was extensive. He may have been a difficult and controversial figure while alive, but after his death his reputation grew to truly significant proportions. His self-proclaimed identity as a reformer led to his being called "the Luther of medicine" by followers and critics alike, after the great religious reformer Martin Luther

(1483–1546). His ideas proved especially popular among Protestants in Germany and France for a number of reasons. Some embraced Paracelsianism because of its reformist tendencies, seeing in its challenge to traditional Galenic medicine a secular parallel to the religious transformations of the Protestant Reformation. Some Protestants in Germany, motivated by nationalist pride, celebrated the fact that Paracelsus had deliberately eschewed Latin and instead had written his medical works in Swiss-German. There is also an interesting connection between the Paracelsian emphasis on seeking out the hidden mysteries of nature and Protestant beliefs that the individual should cultivate a spiritual and moral connection with God. For Paracelsus, the true physician read the Book of Nature in which God had inscribed both wisdom and knowledge; he framed the study of nature as a spiritual and theological exercise that brought the physician closer to God, an idea that appealed to many Protestants in the latter decades of the sixteenth century.

The Paracelsian focus on chemical medicines and the art of alchemy both had a widespread and lasting influence on later medical practitioners. Increasing numbers of physicians embraced chemical medicine from the late sixteenth century onward, leading to serious, sometimes violent disputes between Paracelsians and Galenists in towns and universities across Europe. Proponents of Paracelsian medicine became so numerous that they established their own institutions; one example sprang up in London in the middle of the seventeenth century, when a powerful collective of chemical physicians formed their own college in direct opposition to the Royal College of Physicians and its support for traditional medical practices. Not all chemical medicine was strictly Paracelsian, but its rise can and should be seen as a direct consequence of Paracelsus's work.

Before we leave Paracelsus it might be interesting to compare him with John Dee, the English magus from Chapter 1. These were two men who embraced the study of the hidden parts of nature, though for different reasons and drawing upon different traditions. Together, they provide a useful sense of how magic was understood in this period as well as how different thinkers could hold very different ideas of what magic was and how it was useful.

Dee was rooted firmly in the tradition of learned magic. He was well-acquainted with the Hermetic corpus as well as the cabalistic arts, and he saw the practice of magic as fundamentally concerned with unlocking the hidden mysteries of the universe. Dee was also extraordinarily ambitious: he wanted to unify the warring factions of Christianity by revealing the universal truths of the *prisca theologia*, the ancient theology that had been obscured and lost beneath thousands of years of human history, but he also wanted to use his knowledge of nature to secure greater power for the English throne. Magic was a powerful means to an end for Dee, a tool that could reshape the world in profound ways. At the same time, it had strong connections to his religious faith; magic offered a means of speaking with angels and other spiritual beings, providing access to a critical source of knowledge. Dee's ideas about magic were colored by the fact that he was a Neoplatonist, meaning that he believed that true wisdom lay not in the tangible world around us but in the unseen truths that lay behind and above that world. The hermetic philosophy, which had Neoplatonist elements of its own, was uniquely suited to Dee's worldview.

Paracelsus, however, had a markedly different perspective on both the world and the purpose of magic. He was not a humanist like Dee, with an interest in classical philosophies. He may have had some contact with the Hermetic tradition, but if so his knowledge of hermeticism was less focused on the philosophical and theological ideas embraced by Ficino and more on hermetic traditions that described alchemical processes or talismanic magic in which the magus inscribes symbols on physical objects in an effort to imbue them with natural virtues and powers. Unlike Dee, Paracelsus certainly did not believe that the tangible, mundane world held few sources of wisdom; on the contrary, he believed that the secrets of nature were spelled out in nature itself, left there by a benevolent and providential God. His ambitions were both more limited and more pragmatic than those of Dee: Paracelsus wrote often of reforming the practice of medicine, but more than anything else he wanted to heal the injured and cure the sick. For him, the study of nature was also the study of its hidden qualities and forces; the two were indistinguishable from one another. Thus, both Paracelsus and Dee sought to uncover and harness nature's secret mysteries, but while Dee looked to the abstract arts of mathematics and theology for the key to those mysteries, Paracelsus counseled his readers to use their eyes and hands to investigate the world around them.

While we can understand Dee as a practitioner of learned magic, with its roots in classical antiquity and its erudite commentaries on hermeticism and cabalism, it is more complicated to assign Paracelsus to a particular tradition. In part this is because Paracelsus was an iconoclast, meaning that he was unique; his ideas were very much his own rather than descending from a larger school of thought. We have already seen that he had little patience for the dusty scribblings of antiquity, going so far as to throw Avicenna's *Canon* into a fire in front of his students, and in this respect he was certainly very different from Dee, who had a deep reverence and respect for the philosophies of the ancients. Paracelsus taught that true wisdom would be found not in the pages of a literal book, but in nature itself, the Book of Nature, in which God had inscribed wisdom that waited for those willing to seek them out. Thanks to his wanderings across Europe, he was acquainted with and willing to borrow ideas and practices from folk medicine, which he saw as more closely connected to the hidden secrets of the natural world.

In other respects, however, Paracelsus and Dee shared important similarities. Religion and piety were both central to how these men understood the world and their own efforts to affect it. Dee devoted many years to scrying for angelic messages, hoping to receive the wisdom he needed to reunite the Christian faith and secure the imperial ambitions of the English throne. Paracelsus taught his followers to seek divine messages in the plants and stones of the world around them, and to revere God above all else. Both men saw great power and possibilities in the hidden secrets of nature, and both believed that those secrets held the key to humankind's future. Dee was only fourteen years old when Paracelsus died, but together they illustrate how differently people could connect together the realms of natural philosophy, religion, and magic.

Sympathy and the Devil

One of my favorite examples of a medicine that we might consider "magical" is something called the weapon salve or powder of sympathy. Its first appearance seems to have been in the revised, 1589 edition of Giambattista della Porta's (1535–1615) *Magia naturalis* or "Natural Magic." There, Della Porta attributed the weapon salve to Paracelsus, though it is unlikely that he had anything to do with it. In its earliest descriptions, the weapon salve was a kind of ointment that was able to heal wounds quickly, painlessly, and without infection when it was applied to the weapon that had caused the wound or to traces of the patient's blood. If someone was injured during a duel, for example, his wound could be healed just by smearing the sword that had cut him, even if he was a great distance away. Proponents of the salve believed that there was some manner of *sympathetic* connection between the blood on the weapon and the blood still in the patient, and that the healing virtue of the salve would be communicated from one to the other even across distances of many miles.

This idea of a sympathy or correspondence between two different objects lay at the heart of many magical philosophies in this period, one reason why we might consider this remedy to be, at least in some respects, magical. In fact, sympathetic magic is one of the oldest and most common forms of magic in European history. The idea that what was done to one object could influence another – the very essence of sympathetic magic – appears in learned treatises and descriptions of folk magic alike across many hundreds of years. Such practices were described and recorded in ancient Greece and the Roman Empire, perhaps best exemplified in the *Natural History* of Pliny the Elder (c. 23–79), a huge and comprehensive survey of natural knowledge that contained many examples of sympathetic magic, such as the use of a frog's tongue to elicit the truth from a sleeping person. Pliny himself appeared skeptical of some of these claims, but he recorded them nonetheless. That the weapon salve descended from this long tradition of sympathetic magic is clear. Moreover, the prevalence of sympathetic magic in premodern European folk medicine has led the historian Roberto Poma to describe the salve's appearance in learned treatises as "the social rise of a peasant remedy."[2]

From its first appearance in the 1580s, recipes describing how to concoct and use the weapon salve became increasingly elaborate. Common ingredients were animal (or human) fat, powdered earthworms, *mumia* – pieces of preserved human flesh, known as "mummy" in English – and *usnea*, a kind of moss or lichen scraped from human bones, preferably the skull. If you think those ingredients sound a little creepy, you're not alone. While mummy and *usnea* were both used in numerous medical remedies at this time, for some people

[2] Roberto Poma, "L'onguent armaire entre science et folklore médical: Pour une épistémologie historique du concept de guérison," *Archives de Philosophie*, 73 (2010), p. 603.

living in the sixteenth and seventeenth centuries it all sounded a bit too much like the stereotypical practices associated with witchcraft, especially because the salve also required the addition of human blood. In fact, some people were so concerned that they wrote entire books in which they claimed that the salve was the work of the Devil. A good example is the work of William Foster (b. 1591), an obscure clergyman based in rural England who, in 1631, published a book attacking not just the salve but also one of its most well-known supporters, the English physician Robert Fludd, who we encountered back in Chapter 1. "This stinking Weapon-Salve," wrote Foster, "this Magicall oyntment," was the product of such "diabolicall Conjurations" and "superstitious Operations" that "it cannot be lawfull for an honest and religious man to use it."[3] He went on to accuse Fludd himself of witchcraft and of consorting with the Devil, claims that Fludd was forced to refute in a work he published that same year.

Fludd was a committed occultist, and he integrated medicine, theology, and magic in ways that some people at the time found problematic. His encounter with the sharp pen of William Foster shows us how slippery the slope was when it came to extraordinary or magical claims. Fludd was well known at the time and his works widely read outside of England, and he was connected publicly to occult societies like the Rosicrucians or the Fraternity of the Rosy Cross. Fludd vigorously denied that he was a Rosicrucian, but despite his efforts the label stuck and he was seen by many as a credulous aficionado of all things mystical and magical. Foster's attack, and specifically his claim that Fludd was a devil-worshipping sorcerer, was therefore not a minor problem; in the 1630s, accusations like that sometimes led people into the clutches of the judiciary for the suspected crime of witchcraft.

While Fludd had to contend only with the aspersions cast by a single individual, others were less fortunate. In the decade before his literary encounter with William Foster, the Flemish physician Jan Baptista van Helmont (1580–1644) ran into significant trouble after circulating his own defense of the weapon salve. Van Helmont was a Paracelsian and, in 1621, wrote a defense of what he called "the magnetic cure of wounds" that doubled as a scathing attack on a Catholic priest named

---

[3] William Foster, *Hoplocrisma-spongus: or, A sponge to wipe away the weapon-salve A treatise, wherein is proved, that the cure late-taken up amongst us, by applying the salve to the weapon, is magicall and unlawfull* (London, 1631).

Jean Roberti (1569–1651). Roberti had spent the previous fifteen years sparring with another Paracelsian physician, Rudolph Goclenius the Younger (1572–1621), a professor at the Protestant university at Marburg, which was known from the late sixteenth century as a center for alchemical and Paracelsian studies. Van Helmont took exception to Roberti's attacks against the weapon salve and made a series of aggressive remarks aimed both at Roberti himself and at the Jesuit order, the society to which Roberti belonged within the Catholic Church. He apparently thought himself safe in doing so, as he intended to circulate his work only among his close acquaintances. According to Van Helmont, however, his treatise on the magnetic cure of wounds was published without his permission by Roberti's brother, at which point he found himself in a great deal of trouble. Living as he did in the Spanish-controlled Low Countries (the modern-day Netherlands), his pointed jabs at the Jesuits were viewed in an unfavorable light by the Catholic establishment. Even more troubling was the comparison Van Helmont drew between the weapon salve and the miracles performed through holy relics, both of which purportedly healed the sick without direct contact. He eventually found himself on trial before the Spanish Inquisition, suspected of heresy, and then spent the rest of his life under house arrest.

Van Helmont's unfortunate fate illustrates the way in which particular phenomena could blur the lines between legitimate and natural activity on the one hand and suspect, illegitimate activity on the other. Moreover, his unwise comparison between the mysterious power of the weapon salve and the miracles wrought by God through sacred relics represented a challenge to Catholic doctrine that invited accusations of heresy. What he and others insisted was natural, others could and did construe as demonic or otherwise problematic. As Van Helmont's case demonstrates, these were not idle concerns or empty arguments; for premodern people, they touched on crucial questions about the world.

In the 1640s, perhaps as a result of the controversies that swirled around the weapon salve, it gradually was replaced in learned treatises by references to something called the powder of sympathy, which was basically the same thing but with most of the objectionable bits stripped away. While the weapon salve had invited comparisons to witchcraft with its skull-moss and preserved human flesh, the powder of sympathy appealed to an entirely different audience by casting itself essentially as a chemical remedy in line with other, similar remedies supported by growing numbers of Paracelsian practitioners.

Proponents often described it as a form of vitriol, a volatile chemical compound, and like the weapon salve it was applied to traces of blood and then worked a cure over some distance.

Interestingly, though the mechanism of application and the effects of the cure remained relatively unchanged, explanations for how these different remedies worked altered significantly. Earlier theories for the weapon salve tended to invoke invisible forces similar to magnetism, as in Van Helmont's "magnetic cure of wounds," or the astral and celestial emanations theorized by Goclenius the Younger that conveyed the salve's power across vast distances. These explanations aligned with beliefs about magnetism and astrological magic that were popular around the beginning of the seventeenth century, but by the 1640s these ideas had become less influential. That is when the powder of sympathy first appeared, and its advocates proposed a different kind of explanation for how it was able to heal over distances. They wrote of volatile particles moving through the air, following a path back to the wounded patient that was guided not by some mysterious and intangible sympathy but by linked chains of bloody atoms or the wafting but subtle heat of the open wound. In other words, these explanations were material in nature, something that will make more sense when we reach Chapter 5 and discuss the new mechanical philosophies of nature that became popular in the 1630s and 1640s.

This transformation from a problematic, quasi-magical ointment to a respectable chemical preparation allowed these sympathetic remedies to persist right through the end of the seventeenth century. They were discussed and debated by just about everyone of note, as well as in dozens of works written and circulated by lesser-known academics and physicians. Ultimately, however, belief in remedies that could heal over distances faded away in the early eighteenth century, for reasons that remain unclear. Perhaps their amazing healing powers could not survive the rigorous and empirical approach of the "new science" that emerged in the latter half of the seventeenth century, or perhaps they simply fell out of fashion. Their longevity and adaptability, however, show us that even something seemingly "magical," powered by occult and mysterious forces, could evolve in parallel with changing philosophical explanations until it shed its magical connotations altogether. This in turn provides a hint as to the ultimate fate of many other magical ideas in this period.

4 A New Cosmos

Copernicus, Galileo, and the Motion of the Earth

In 2006, the members of the International Astronomical Union (IAU) voted to change the long-standing definition of the term "planet." Before this, a planet was an object that orbited the Sun and was large enough to have assumed a nearly spherical shape, and under that definition our solar system was thought to have nine planets. Beginning in the late twentieth century, however, astronomers began to discover a series of bodies at the far reaches of the solar system that appeared to be roughly the same size as tiny Pluto, or even bigger. Were all of these now to be classified as planets? That seemed problematic. Rather than embracing these other bodies as planets, the members of the IAU voted to add another criterion to the existing definition: to be classified as a planet, a body also needed to possess enough gravitational force to "clear the neighborhood" surrounding its orbit, meaning that no bodies of similar size could lie close to its orbital path. With that one vote, Pluto was no longer the ninth planet from the Sun. Instead, it joined the growing numbers of "dwarf planets" that orbited the Sun at the outer edges of the solar system.

This change in Pluto's status caused a global outcry. Tens of thousands of people signed petitions urging the IAU to reconsider, and some astronomers were equally unhappy for technical and scientific reasons. Members of the California state legislature actually put forward a tongue-in-cheek resolution that called the demotion of Pluto "a hasty, ill-considered scientific heresy similar to questioning the Copernican theory" and a decision that "renders millions of text books, museum displays, and children's refrigerator art projects obsolete."[1] Even now, more than a decade after Pluto's change of status, some people refuse to accept it.

---

[1] Edna DeVore, "Planetary Politics: Protecting Pluto," *Space.com* (September 07, 2006). www.space.com/2855-planetary-politics-protecting-pluto.html.

All this fuss over a distant piece of rock! After all, astronomers didn't announce that what they thought was Pluto was actually just a smudge on their telescopes or that their math was wrong and, whoops, Pluto is really Earth's second Moon. Our solar system was the same after the IAU's vote as it had been before. And yet, as the public reaction to Pluto's fate might suggest, even so minor a change in definition was enough to cause worldwide confusion and dismay.

Now imagine that the president of the IAU had stepped up to a microphone in 2006 and told the world that the *entire universe* was completely different from what every astronomer, scientist, and third-grade teacher had taught us – that, in fact, almost everything we thought we knew about the universe was wrong.

It is difficult to imagine that scenario because we live in a time when our understanding of the universe has remained fairly constant, supported by ever-increasing amounts of data and evidence. While scientists continue to make new discoveries like the Higgs boson or mysterious dark energy, most of us believe that we understand at least the fundamental facts about our universe. People living in sixteenth-century Europe thought the same thing. We now know that they were wrong.

In this chapter, we explore how the European understanding of the cosmos changed in the sixteenth and seventeenth centuries. It was one of the single greatest intellectual disruptions in European history, and in some ways we are still feeling its effects now, more than 450 years later. The claim that our universe was fundamentally different from what people had known for thousands of years led to a serious conflict between different sources of knowledge and forms of authority, and forced premodern Europeans to grapple with a crucial question: Who has the right to define the nature of reality?

This particular conflict is often framed by historians and other commentators as a battle between science and religion in which the brave and progressive pioneers of the heliocentric cosmos were attacked unjustly by a tyrannical and old-fashioned Church. This is an exaggeration, but not by much. Until the latter decades of the twentieth century, most of the histories written about this time were skewed heavily in favor of "science." There are good reasons for this. The Catholic Church did condemn the idea that the Earth moved, and eventually placed the famed mathematician and astronomer Galileo Galilei (1564–1642) on trial for heresy. It was only in 1992, more than

350 years later, that Pope John Paul II (1920–2005) admitted that the persecution of Galileo had been unjust. It is indisputable that the Catholic Church, and religion more broadly, were important reasons why premodern Europeans were so reluctant to change their perspective on the cosmos.

If we look at these events from the perspective of the Church, however, we gain a somewhat different view. Theologians and clergy in the Catholic Church believed that one of their most important tasks was to help the average person understand both the world around them and their place in that world. When Galileo and others began to argue for a new and different vision of the world, many theologians saw this as a serious problem with ramifications that stretched far beyond the realm of natural philosophy. Thus, the story told in this chapter is actually more complicated and interesting than a fight between "science" and "religion." The tensions described here played out in the words and thoughts of people trying to reconcile an older, familiar cosmos with a series of new discoveries and ideas that were literally earth-shaking in what they suggested about the universe and humanity's place in it. There are no heroes and villains in this story, only the slow-moving collision between two different worlds.

The Premodern Cosmos

Questions about the nature of the universe were the domain of natural philosophy, and for premodern Europeans the single most important natural philosopher was the Greek scholar Aristotle (384–322 BCE). His works had been translated and commented upon for centuries in the Roman Empire and then the Islamic world, along with other examples of Greek natural philosophy that included works by Plato, Democritus (c. 460–c. 370 BCE), and Epicurus (341–270 BCE). The sheer volume of Aristotelian philosophy that survived and passed into the Latin West, however, made Aristotle preeminent; medieval authors regarded him so highly that they referred to him simply as "the Philosopher."

Aristotelian natural philosophy centered around questions of *why* objects exist and sought to understand the causes for phenomena and behaviors that we observe in nature. Many people who came after Aristotle found his philosophy appealing because they too wanted to understand why and how things happened. But Aristotelian philosophy was also *teleological*, which means it embraced the concept of purpose (the Greek word *telos* means "goal"). For Aristotelians, everything in the world had a purpose or goal toward which it progressed, changed, and moved. That purpose was not assigned by a higher power, however; Aristotle wrote virtually nothing about the gods. It was simply innate, part of each different thing.

Aristotelian philosophy was taught almost continuously for close to 2,000 years, and while it evolved and changed over time as others interpreted and expanded it, many of its fundamental principles remained largely intact. In part, this is because Aristotelianism is fairly simple as philosophies go. Aristotle believed that the philosopher should study examples of change (or, as he called it, "motion") in the natural world, and he taught that change is the result of basic and immutable properties possessed by all natural substances. A heavy object will always fall toward the center of the Earth because it is trying to move toward its natural place; what that place is and how the object moves toward it are both dictated by the innate properties of the object itself. At times, Aristotelian philosophy can appear *tautological*, meaning that it contains a form of redundant reasoning or phrasing; for example, Aristotle argued that a heavy object falls to the ground because it is heavy. It may not be a very satisfying explanation, but it is certainly easy to understand.

The wider universe described by Aristotle was fairly simple (Figure 4.1). The Aristotelian cosmos was divided into two parts: the *terrestrial* realm (which included the Earth itself, immobile at the center of the universe) and the *celestial* realm (which included the seven known planets and the fixed stars). The division between these realms fell along the orbit of the Moon, the "planet" closest to the Earth, which is why the terrestrial realm was also called the *sublunary* realm ("below the Moon") and the celestial was known as the *superlunary* realm ("above the Moon"). Everything in the terrestrial or sublunary realm was composed of the four classical elements – earth, water, air, and fire – and was subject to continuous change. Beyond the Moon's orbit was the celestial realm, which was composed of a single, fifth element called *quintessence*. Unlike the terrestrial realm, which was in a constant and turbulent process of change, the celestial realm was eternal and unchanging. There, the seven planets known to the ancient Greeks – the Moon, Mercury, Venus, the Sun, Mars, Jupiter, and Saturn – orbited the Earth in perfect circles. Beyond the orbit of

Figure 4.1 A depiction of the geocentric cosmos, with the four terrestrial elements at the center surrounded by the orbits of the seven planets and the sphere of fixed stars. From Peter Apian, *Cosmographia*, 1539.
Photo by Photo12/Universal Images Group via Getty Images

Saturn was the sphere of fixed stars, which marked the outermost boundary of the cosmos and which also rotated, carrying the stars around the Earth.

This was the cosmos generally accepted and understood by Europeans until the sixteenth century. To modern eyes it might appear to be a very small universe that is also fundamentally wrong, but there were good reasons why premodern people believed this was the basic structure of their world. For example, the idea that the Earth is immobile at the center of the cosmos makes sense when, as far as anyone can tell, it does not move. Consider what your own senses and experiences tell you. Does it feel like you are being carried around the Earth's axis at anywhere from 700 to 1,000 miles an hour while *at the same time* hurtling around the Sun at 67,000 miles an hour? The simple answer is, no. Everything we can see and feel tells us that we live on an object that is not moving. Instead, what we *do* see moving are the Sun, the Moon, the stars, and the other planets. Therefore, if we base our understanding of the world on our own experiences, we should conclude that the Earth does not move while everything else in the universe rotates around it.

This kind of thinking was particularly attractive to Aristotle, who valued direct experience. His cosmology – that is, his description of the physical universe – not only coincided with prevailing ideas already present in Greek philosophy, such as the four elements, but was also reinforced by the daily experience of everyone living on Earth. This is why Aristotle's philosophy seems to make common sense, which in turn helps to explain its popularity across many hundreds of years. His understanding of the world was based on simple and obvious ideas, such as that the Earth is immobile, or that nature is defined by change or motion. We only have to look around ourselves to see that we live on a planet where nothing remains static or unchanging: cold things warm up, water evaporates, plants and animals and people grow and die and decay, mountains rise up and are worn down. Change is constant, which is why constructing a natural philosophy around the concept of change or motion makes a great deal of sense. In turn, this is a powerful reason why Aristotelian philosophy reigned supreme in premodern Europe.

Astronomy before Copernicus

Today we usually think of astronomy as a field of science that both observes the wider universe and explains the origins and behavior of celestial objects. Historically, however, astronomy was little more than applied mathematics, like geometry or arithmetic. In antiquity, astronomers observed the heavens and used those observations to predict events like eclipses, the summer and winter solstices, and planetary conjunctions. Before the existence of a standardized calendar the heavens were the best and most accurate way to track time, and many ancient civilizations used the movements of the planets to mark ritualistic occasions, explain past events, and foretell future calamities. As a result, the ability to predict celestial events and motions was very important.

Astronomers, then, were mathematicians who observed the heavens and offered predictions about the movements of stars and planets. They did not usually speculate about the natures of those objects or propose theories that explained why or how they moved. Those kinds of questions were philosophical rather than mathematical, and while the two realms might overlap – some astronomers did in fact speculate as to how the heavens moved, while some philosophers, like Aristotle, used astronomical observations to support their claims about the physical nature of the cosmos – for the most part they were understood to be separate endeavors. This is important because in the sixteenth and seventeenth centuries a number of people, foremost among them Galileo Galilei, began to argue that this no longer should be the case. That, in turn, led directly to Galileo's infamous encounter with the Catholic Church.

While many ancient astronomers focused on predicting the motions of the heavens instead of explaining them, some also proposed purely theoretical models that aligned with what they saw in the skies. These models were not necessarily intended to be explanatory in a philosophical sense – that is, they did not have to establish the physical causes for planetary motion nor speculate, for example, as to what substances composed the planets themselves. Instead, these models were used to make predictions easier and more consistent. Creating theories or hypotheses about the heavens to account for existing observations is an example of what historians sometimes call "saving the phenomena," a way of explaining the celestial realm in which any theoretical model was held to be useful so long as it matched with, or "saved," the phenomena observed in the skies. According to this way of thinking, an astronomical model need not be physically "true" or "real" to be useful to the astronomer; it need only reflect and account for what they observed. This made sense to premodern people because astronomy and natural philosophy were seen as separate disciplines concerned with answering different kinds of questions.

A good example of a theory "saving the phenomena" was proposed several hundred years after Aristotle by Claudius Ptolemy (c. 100–170), a Greek mathematician and astronomer who lived in Roman-controlled Egypt. The Ptolemaic system, as we now know it, was outlined in a text called the *Almagest*, which gathered together centuries of ancient understanding about the universe and remains today the only surviving work of classical antiquity dedicated entirely to astronomy (the ideas about the cosmos held by Aristotle and others were recorded only as part of larger works on natural philosophy). Ptolemy's vision of the cosmos did provide a coherent theory of planetary motions, but it was intended more to align with astronomical observations than to reflect the physical reality of the heavens.

Along with many other astronomers, Ptolemy wanted to devise a means of addressing a particular quirk of planetary motions called "apparent retrograde motion." If someone goes outside night after night and observes the path taken by Mars through the sky, over time they will see it moving steadily in one direction against the backdrop of the more distant stars. Periodically, however, Mars appears to slow down, come to a stop, and then reverse its direction for a while, before stopping again and then resuming its original motion. This retrograde or backwards motion is also exhibited by other planets in our solar system when observed from Earth. Today we understand that this behavior is caused by the different speeds at which Earth and the other planets orbit the Sun: because the Earth moves around the Sun more quickly than Mars, eventually we catch up to Mars and then pass it, and when that happens it appears to someone standing on Earth that Mars actually moves backwards for a time. But to the ancients, who lived in what they thought was a geocentric universe, this behavior was difficult to explain.

To save the phenomenon – in other words, to create a theoretical structure for the cosmos that explained this observation – ancient astronomers came up with some fascinating and complicated ideas. Ptolemy suggested that these and other planetary motions could be understood using a complex system of geometrical operations in which each planet moved simultaneously along two different paths. The first was called the *deferent*, a perfect circle centered on a point called the *eccentric* that sat at the middle of the universe. The Earth was offset from the eccentric just slightly, an adjustment that allowed Ptolemy to account for the fact that the seasons were not of equal length, meaning that the Sun was slightly closer to the Earth at one point during the year and further away at another point. Attached to and moving around each deferent was another perfect circle called an *epicycle*, and it was around the epicycle that each planet moved. To summarize, then, the deferent is a circular path around the Earth, the epicycle is

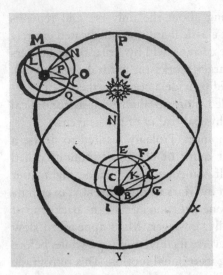

Figure 4.2 A 1643 woodcut depicting epicycles (smaller circles) and deferents (larger circles). On one, Earth and Moon orbit the Sun; on the other, the Sun and Moon both orbit the Earth.

another circle moving around the deferent, and each planet moves around its epicycle (Figure 4.2). While complicated, the beauty of Ptolemy's system is that it provided a geometrical model that accounted for apparent retrograde motion; if someone were to track the motion of a planet around both deferent and epicycle, it would follow a path that, from the perspective of the Earth, exhibited the retrograde motion we observe. Thus, the Ptolemaic system is probably the best-known example of a theory that attempted to "save the phenomena" by proposing a purely mathematical model that predicted the complicated motions of the planets as observed from Earth.

The Ptolemaic system survived and flourished for almost 1,500 years. It was used, examined, and critiqued by many thousands of people over that time, and it persisted because it worked. It allowed astronomers to predict and measure celestial motions at the same time that it offered a mathematical model that made those predictions possible. It was plausible enough to be useful without necessarily explaining how the physical universe actually looked or behaved. This is important: Ptolemy's deferents and epicycles were not necessarily "real" in a physical sense. They were theoretical, and while they *could be* real, they did not need to be in order for astronomers to predict and calculate planetary motions.

This is where the history of astronomy becomes complicated, because alongside Ptolemy's theoretical system were other ways of understanding the universe that claimed to describe actual, physical reality. Long before Ptolemy, Aristotle proposed a cosmology that demanded explanations that were anchored not in plausible and abstract theories but in the physical mechanisms and motions that operated in the universe. Whereas Ptolemy wanted theories that would allow the astronomer to track and predict the motions of the heavens, Aristotle wanted to explain what *caused* that motion. He proposed that the Earth was surrounded by a series of crystalline spheres, each inside the next, and that the planets moved around the Earth within these spheres. Remember that all of this was made from quintessence, the fifth element, and that everything moved in perfect circles because the heavens were perfect and unchanging. When Aristotle taught his students about this geocentric cosmos, he believed that he was describing what actually existed in the skies.

Aristotle's vision of the cosmos was only one of many that existed in classical antiquity, but unlike most of its competitors it survived into the sixteenth and seventeenth centuries. This means that early modern Europeans inherited two different cosmological systems from classical antiquity: Ptolemaic and Aristotelian, each describing the cosmos but devised for different purposes and contradictory in fundamental ways. This is important because these two systems together created a division between *abstract theory* and *physical reality*. This meant that people in premodern Europe were not only capable of understanding their universe in profoundly different ways at the same time, but also saw these two competing methodologies as a necessary division between two different kinds of work: either producing mathematical predictions of planetary motions or explaining the physical nature of the universe. It was not as simple as different people falling into different camps, either Ptolemaic or Aristotelian, because the same person might use each system to do different kinds of things. It also was not the case that these two systems did not overlap, because they did. This division between two different models of the universe, however, would have an enormous impact on the astronomical debates that took place in the sixteenth and seventeenth centuries.

Figure 4.3 Portrait of Nicolaus Copernicus, c. 1515.
Photo by Hulton Archive/Getty Images

The Copernican Revolution

In 1543, a book appeared in the city of Nuremberg that changed how some Europeans looked at the universe: *De revolutionibus orbium coelestium* or "On the Revolutions of the Heavenly Spheres." Its author was Nicolaus Copernicus (1473–1543), a Polish astronomer and mathematician (Figure 4.3). Copernicus had strong ties to the Catholic Church; he was a canon, which meant he was responsible for maintaining a cathedral (the seat of a bishop or archbishop), and some historians believe that he was ordained as a priest as well. There is an irony here, for Copernicus's work would be used to challenge some important teachings of the Catholic Church long after the man himself died.

We know that Copernicus began working on what would become the *De revolutionibus* as early as 1513 or 1514, several decades before it finally appeared in print. Most of it probably was completed by the mid-1530s, but Copernicus was reluctant to publish it right away because his work called into question some of the most fundamental assumptions about the universe held at the time. He did decide to circulate his ideas quietly among other astronomers, however, and after seeing that his calculations were not rejected outright Copernicus finally had his work printed in Nuremberg shortly before his death. It wasn't exactly a bestseller. The *De revolutionibus* was an extremely technical piece of work with a lot of complicated mathematics, which might explain why sales of its first print run were relatively small. This led the twentieth-century historian Arthur Koestler to dub *De revolutionibus* "the book that nobody read," though later historians, particularly Owen Gingerich, established that the book was in fact owned by most of the period's prominent astronomers and mathematicians. Just owning a book, however, does not necessarily mean that someone agrees with it, or has even read it.

The universe described by Copernicus in *De revolutionibus* would have looked familiar to most people at the time, but with one or two crucial differences. Most famously, Copernicus swapped the positions of the Earth and the Sun, making the Copernican system heliocentric rather than geocentric. The Moon continued to orbit the Earth, but the six planets of the known universe – Mercury, Venus, Earth, Mars, Jupiter, and Saturn – now revolved around the Sun (Figure 4.4). The apparent movement of the Sun across the sky, as well as the movements of the stars at night, were caused by the Earth rotating on its axis. Thus, the Copernican system introduced the idea that the Earth experienced two different kinds of motion simultaneously: rotation around its axis and rotation around the Sun. As innovative as this was, however, Copernicus also preserved some features of the standard astronomical models that preceded his theory. For example, he argued that the planets moved in perfect concentric circles just as Aristotle had taught, and also advocated for the use of theoretical constructs (including Ptolemaic deferents and epicycles) to aid astronomers in calculating the positions and motions of the planets. Ironically, he had to retain Ptolemaic epicycles because of his belief that the planets moved in perfect circles. His insistence on circular orbits conflicted with more recent astronomical observations, meaning that to "save the phenomena" Copernicus had to keep some Ptolemaic innovations for his own system to make the math work. It would be more than seventy years before Johannes Kepler (1571–1630) presented the idea that the planets moved in elliptical rather than circular paths, which finally allowed astronomers to discard epicycles and deferents once and for all.

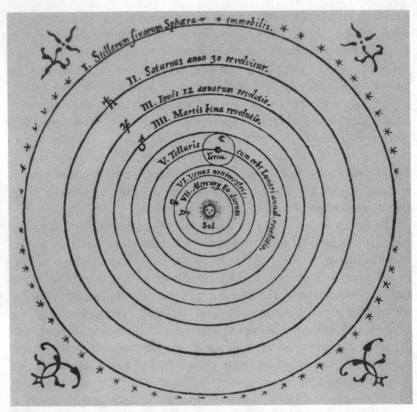

Figure 4.4 A sixteenth-century depiction of the Copernican cosmos with the Sun at its center.

Photo by Bettmann/Getty Images

While the Copernican system represented a significant break from older astronomical theories in some respects, in other respects it left those older theories relatively untouched. Recall that Andreas Vesalius, whose *Fabrica* appeared in the same year as *De revolutionibus*, also retained some ancient ideas about human anatomy even as he challenged and discarded many others. Both of these works demonstrate how difficult it was for early modern Europeans to overthrow completely the ideas of classical antiquity. Nevertheless, Copernicus set out to create a more precise and elegant astronomical description of the movements of the heavens than what already existed, and in this he succeeded. He claimed in his dedication to *De revolutionibus* (addressed, it should be noted, to Pope Paul III) that thousands of years of astronomical calculations had not produced an adequate theory of celestial motions, and his hope was that his own work would be a first step toward such a theory. The real genius of Nicolaus Copernicus, however, is that he devised his heliocentric theory using little more than mathematics. He did not have a telescope with which he could gather new observations, as Galileo did some seventy years later. Instead, he had a series of observations made with the naked eye, the astronomical tables produced by Ptolemaic models, and a lot of math.

De revolutionibus was not a bestseller and, according to the research done by Owen Gingerich, those who did buy copies tended to ignore the cosmological theories at the beginning of the work (what most of us today would consider the really important ideas about a heliocentric cosmos) and focused instead on the complicated mathematics that Copernicus used to support his claims. This is one reason why *De revolutionibus* failed to excite much controversy until some time after it was published; most people who read it skipped over the controversial information at the beginning. Other astronomers of the time also took great pains to present the work as a piece of mathematical theory rather than a description of physical reality, probably because they understood what would happen to the Copernican model if it claimed to describe the "real" universe. In fact, Copernicus himself encouraged this practice when he argued in the preface of *De revolutionibus* that it was primarily a work of mathematics rather than of physics or natural philosophy. Historians remain divided as to why Copernicus did this; it is possible that he, too, anticipated serious challenges to his theories if they were seen as describing the physics of the cosmos rather than providing a mathematical model of its motions. It is also possible that Copernicus himself was uncertain as to whether the universe was actually heliocentric, given that his stated goal in writing *De revolutionibus* was to provide a more precise and elegant means of charting and predicting celestial motions rather than to present a new cosmology.

Here we see, again, this division between abstract theory and physical reality, a division of which Copernicus obviously was aware. It was only exaggerated by one of the most interesting and famous aspects of *De revolutionibus*, which was something added to the beginning of the work not by Copernicus himself but by a Lutheran theologian named Andreas Osiander (1498–1552), who oversaw the publication of *De revolutionibus* but was concerned by the theological implications of the Copernican theory. Before it went to print he added an unsigned preface to the beginning of the work, something that Copernicus probably knew nothing about but which many people originally attributed to him.

Osiander's preface did something strange by modern standards, but also very important. It presented the heliocentric Copernican universe as merely *possible* rather than probable or "real." According to Osiander, the cosmology suggested by Copernicus was useful only in calculating and predicting astronomical events; it was a theoretical description of a possible universe, a kind of thought-experiment rather than a description of physical reality. As a theologian, Osiander was worried that Copernicus's ideas might conflict with Christian scripture and wanted to establish a convenient middle ground in which the Copernican system could be published and studied as a mathematical model of a hypothetical universe, leaving aside the trickier questions about the actual reality of the cosmos. It was a kind of compromise that would remain a part of cosmological debates for more than a century.

How Osiander phrased his interpretation of the Copernican model in particular and astronomy in general is worth considering, because it highlights an important idea that he shared with many people in the sixteenth century. Osiander wrote that "it is the duty of an astronomer to compose the history of the celestial motions through careful and expert study. Then he must conceive and devise the causes of these motions or hypotheses about them. Since he cannot in any way attain to the true causes, he will adopt whatever suppositions enable the motions to be computed correctly." Note that Osiander made a sharp distinction between "hypotheses" or "suppositions" on the one hand, and "true causes" on the other. This is important because it points to deeper, fundamental principles of premodern ideas – specifically, ideas about who has the right to establish or proclaim the "true causes" of things. For Osiander, this was not the purpose of astronomy nor the role of the astronomer: "Let no one expect anything certain from astronomy, which cannot furnish it," he wrote, adding that "these hypotheses need not be true nor even probable. On the contrary, if they provide a calculus consistent with the observations, that alone is enough."[2]

---

[2] From Edward Rosen's translation of *De revolutionibus* (Warsaw: Polish Scientific Publications, 1978).

What Osiander suggested in his preface was not controversial. In fact, it aligned with how most educated Europeans understood the distinction between philosophy and mathematics (of which astronomy was a part). Philosophy investigated the true causes of things; it tried to establish why particular behaviors or events happened in the universe as well as fundamental questions like the nature of matter, the behavior of moving bodies, and many other important concepts. Mathematics, however, could not provide answers to such questions. It might describe the trajectory of a moving body, for example, but it had nothing to say about *why* that body moved or *what* composed it. Mathematics, at its core, was the study of abstract hypotheticals; in its purest form it was concerned with imagined scenarios or with universal relationships and proportions, not with the messy ambiguities of the real world. When Osiander argued that astronomy was not concerned with "true causes," he was stating a fact that most of his contemporaries would have seen as obvious and correct. That most of the people who first read *De revolutionibus* seem to have skipped right over the cosmological arguments at the beginning – the more philosophical content, in other words – would appear to confirm that these distinctions were upheld even among astronomers. Moreover, Copernicus took this same line himself when he dedicated his work to Pope Paul III and argued that his heliocentric system could be used to create a more accurate calendar. He did not argue that his system described the way the universe actually was.

We will probably never know what Copernicus believed about his own system. What we do know, however, is that Copernicus himself, theologians like Osiander, and most of the astronomers who read *De revolutionibus* shortly after its publication were all united in presenting this work primarily as a piece of complex mathematics that promised to reform the practice of astronomy and little else beyond that. The Copernican model did not remain hypothetical, however. Eventually, some people began to speculate that the cosmos described in *De revolutionibus* was physically real, and this is where we start to see a division between philosophers and theologians – what many now call a clash between science and religion.

The Copernican Problem

Osiander's careful preface to *De revolutionibus* had one real purpose, and that was to rule out any claim that the Earth actually, physically moved. Astronomers might use the Earth's motion as a convenient fiction or thought-experiment to help them in drawing up their calculations, but that was as far as it went. Some of Copernicus's friends and colleagues, however, such as his pupil Rheticus (1514–74, also known as Georg Joachim de Porris), were infuriated by Osiander's involvement because it effectively silenced any philosophical or cosmological contributions made by the Copernican model. The question of whether the Earth moved was not merely academic, at least for some people in the sixteenth century. If Copernicus had suggested a model to replace the Ptolemaic system, then his suggestion that the Earth moved should be taken seriously.

Whatever some like Rheticus believed, however, there were plenty of reasons to doubt the Copernican theory that the Earth moved around the Sun. The experience of our own senses suggests that we live on a stationary planet, and supporting this were arguments drawn from the Christian scriptures. In fact, both Catholic and Protestant scholars found themselves in a difficult position when it came to the question of whether the Earth moved. On the one hand, most theologians at this time believed that the natural world was the work of God and, when studied, could reveal important truths about the divine plan for Creation. This meant that natural philosophy could and should play a crucial role in interpreting and understanding the works of God. On the other hand, there were numerous passages in Scripture, thought by most to be the revealed word of God, that clearly indicated that the Earth did not move.

This last fact was what drove Martin Luther, the Protestant reformer, to speak out against Copernicus even before *De revolutionibus* was published. He was quoted as saying in 1539, "People gave ear to an upstart astrologer who strove to show that the earth revolves, not the heavens or the firmament, the Sun and the Moon... This fool wishes to reverse the entire science of astronomy; but sacred Scripture tells us that Joshua commanded the Sun to stand still, and not the earth."[3] The Biblical passage to which Luther referred was cited by almost every early modern commentator on the question of the Earth's movement. It is found in the story of Joshua, who asked God to aid the Israelites in battle, after which "the Sun stood still, and the

[3] Quoted in Thomas Kuhn, *The Copernican Revolution* (Cambridge, MA: Harvard University Press, 1957), p. 191.

Moon stayed, until the people had avenged themselves upon their enemies. ... So the Sun stood still in the midst of heaven, and hasted not to go down about a whole day."[4] Nor was the story of Joshua the only place in the Christian scriptures where someone could find proof that the Earth was immobile. For example, 1 Chronicles 16:30 says, "Fear before him, all the earth: the world also shall be stable, that it be not moved." Psalm 96:10 says that "the world also shall be established that it shall not be moved," and Psalm 104:5 describes God as He "who laid the foundations of the earth, that it should not be removed forever." While very few people living in the West today believe that the Earth is immobile and even fewer people use the Bible to argue that point, these particular scriptural passages were extremely important in early modern discussions about the nature of the cosmos.

Astronomy after Copernicus

Though relatively few people seem to have read the *De revolutionibus* in the decades after it was published, it slowly assumed a role of greater importance in discussions about the universe. Nor was Copernicus the only astronomer in the sixteenth century to propose a new model of the cosmos – in 1588, the Danish astronomer Tycho Brahe (1546–1601) published a work in which he described his own cosmological system. Brahe was determined to construct a cosmology that respected Aristotelian physics, and one central idea of those physics is that heavy objects fall toward the center of the universe, meaning for Brahe that the Earth itself must be at that same center. Brahe was also a skilled astronomer, however, and he recognized that a purely geocentric model could not account for the observations that he and others had made. What he created, then, was essentially a compromise between the Aristotelian and Copernican systems. In his cosmology, the Earth remained stationary at the center of the universe, circled by both the Sun and Moon. The other five planets, however, now orbited the Sun, an innovation that allowed Brahe to predict astronomical phenomena with the same exactitude offered by the Copernican system without accepting that the Earth moved.

This is known today as the *Tychonic system*, and it is important to emphasize that Brahe's reasons for presenting this new cosmology

[4] Joshua 10:13.

were rooted in religious as well as philosophical concerns. Like many in sixteenth-century Europe he saw the Christian scriptures as evidence that the Earth did not move. When his protégé Johannes Kepler tried to persuade him to adopt the heliocentrism of the Copernican system, Brahe's response was clear. He could not embrace a cosmology that so obviously conflicted with the Bible. It is not surprising, then, that the Tychonic system was adopted by the Catholic Church in the years following Brahe's death in 1601. Prompted in part by the discoveries made by the troublesome Galileo, the Church needed a cosmological alternative that did not conflict with scripture, and the Tychonic system provided exactly that.

Kepler's ideas are relevant here too because they demonstrate that it was possible for someone to be a religious and pious person while also embracing a heliocentric cosmos. Kepler was born in what is now Germany and he worked with Brahe for a time in Prague, eventually becoming the imperial mathematician in the court of the Holy Roman Emperor Rudolph II after Brahe's death. Unlike Brahe, however, Kepler was absolutely committed to the Copernican model. In 1596, he published his *Mysterium Cosmographicum* or "The Cosmographic Mystery," in which he proposed that the universe had been designed by God in accordance with the principles of geometry. More specifically, Kepler proposed that the distances between the six known planets could be explained by the different ratios of the five Platonic solids. This is a little tricky to explain, though one of the most famous images from the *Mysterium Cosmographicum* will help (Figure 4.5).

Kepler believed that mathematics and, more particularly, geometry were the foundation for the visible universe. He found a way to take the five perfect, regular geometrical shapes first identified by Plato (the tetrahedron, the cube, the octahedron, the dodecahedron, and the icosahedron) and use their ratios to one another to explain why the planets sat at certain distances from the Sun. He imagined that each polyhedron was nestled within a sphere representing a planetary orbit, expanding out from the Sun at greater and greater distances. He did not claim that there actually were enormous cubes or octahedrons floating in space, but he did think that there was a mathematical design to the structure of the universe, one created by God. While he would adjust some aspects of this theory over the rest of his lifetime, Kepler never entirely abandoned the idea that the structure of the universe was

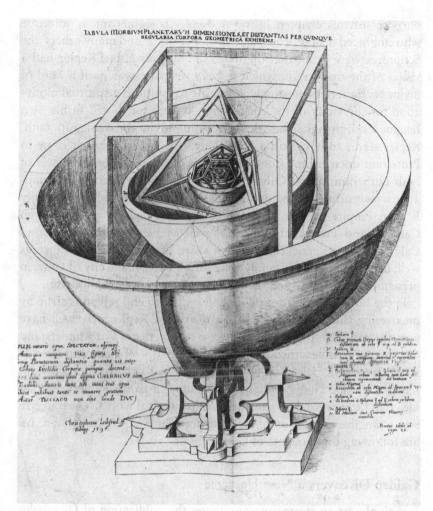

Figure 4.5 Johannes Kepler's model of the solar system, arranged according to the geometrical proportions of the five Platonic solids. From his *Mysterium Cosmographicum*, 1596.

Photo by Hulton Archive/Getty Images

governed by complex mathematical relationships and harmonies designed by God.

There is more that could be said about Kepler, including the fact that he improved upon the work of Copernicus by proposing three laws of planetary motion that are still taught in schools today. For the purposes of this chapter, however, Kepler is significant as someone who embraced heliocentrism *and* faith. For Brahe and many others, the Scriptures were proof that the Earth did not move, but Kepler had a vision of the universe in which its very structure was itself a kind of divine revelation. Interestingly, his desire to establish a spiritual motivation for the study of the cosmos may have been rooted in his own religious misgivings. Though he was raised in the Lutheran faith, Kepler seems to have had a fairly liberal sensibility when it came to Protestant doctrine and questioned some of the fundamental tenets of both Lutheranism and Calvinism, the two main Protestant denominations in central Europe at the time. Living as he did in the Holy Roman Empire, Kepler was surrounded by religious strife between Catholic and Protestant leaders as well as between Lutherans and Calvinists. He experienced his own share of religious intolerance as well: in 1600, he and his family were forced to leave the city of Graz, in present-day Austria, when they refused to convert to Catholicism.

In the same way that John Dee had sought to end religious strife by uncovering the truths of the *prisca theologia*, Kepler may well have seen the rational study of the universe as a better way to understand God than the endless theological squabbles that overshadowed most of Kepler's own life. Historians often attribute to Kepler the quote, "I am merely thinking God's thoughts after Him," a handy summary of an idea that he expressed in many of his astronomical works. The universe was a tangible expression of God's divine plan and, by studying it, we are following the thoughts of God Himself.

Galileo Discovers a New Universe

It took almost an entire generation after the publication of *De revolutionibus* for the ideas of Copernicus to reach a wide audience, but by the beginning of the seventeenth century the Copernican model had acquired significant numbers of followers, Kepler prominent among them. And then everything changed: Galileo Galilei, an Italian mathematician and philosopher, made a series of discoveries that forever changed how Europeans understood the universe. Galileo was born in Pisa, which was then ruled by the powerful Medici family. You might recall that it was Cosimo de' Medici who asked Marsilio Ficino to translate the Hermetic Corpus in the late fifteenth century; now, more than a century later, his descendants would play a pivotal role in

Figure 4.6 Engraving of Galileo Galilei, c. 1640.
Photo by Hulton Archive/Getty Images

another major philosophical discovery. Galileo's father was a noted musician as well as highly educated, and he had big plans for his son. He persuaded the young Galileo to enroll at university to study medicine, though Galileo quickly became fascinated with mathematics. Nine years after first enrolling at the University of Pisa he was appointed the chair of mathematics there (Figure 4.6).

Galileo's achievements are indisputable – he was a talented mathematician as well as a thoughtful philosopher – but, like many other scholars and artists living in the seventeenth century, he found it difficult to earn a living. Many philosophers and naturalists were either independently wealthy or connected to universities, or both. Those who desired greater renown and financial security, however, operated within the early modern institution of *patronage*. This was a system in which a talented individual (a painter, a sculptor, a poet, a mathematician) sought and received the tangible support of a wealthy and powerful patron. Virtually all of the great works of art we know from the Renaissance and early modern period are the result of patronage, produced by artists financed by popes, queens, princes, and others. Patronage did not just supply money, however; it was also a form of social currency. Patronage opened doors to those who might otherwise lack an elevated social standing, and the more powerful your patron, the higher your own standing could become. In many cases, this also meant that you had further to fall when your patron abandoned you, as happened not infrequently in the cutthroat world of early modern social politics.

Universities in the seventeenth century usually paid professors enough to allow them to live in some comfort, but Galileo complained bitterly about his salary as a professor of mathematics first at Pisa and then at the university at Padua. One problem for Galileo was that professors of mathematics were paid significantly less than professors of philosophy, reflecting the lower intellectual status of mathematics at this point in history; while at Padua, for example, Galileo earned substantially less money than his friend Cesare Cremonini (1550–1631), the professor of Aristotelian physics. He tried to secure noble patrons in order to pressure these universities to pay him more money, doing what every client was expected to do in order to gain the patronage of his social betters. He flattered them extravagantly, dedicated works to them, prepared lavish presents, and even tutored their families in mathematics. He soon proved to be extraordinarily shrewd when it came to securing support for himself and his work.

What changed everything for Galileo was the invention of the telescope. Unfortunately, it is not clear who actually invented it. Credit is usually given to a German lens grinder named Hans Lippershey (1570–1619) who filed a patent for an early telescope in 1608, though others working at around the same time seem to have had a similar idea. These early telescopes were simple devices that used two lenses sitting a certain distance apart within a leather tube, and they were capable of somewhere around 3x magnification, a feat that was revolutionary for the time with obvious applications for the military, seafaring vessels, and civil defense. Some philosophers, however, quickly realized that this new technology could have powerful implications for astronomy. One such individual was Galileo.

Galileo lived in Venice when the first telescope arrived there in 1609. Very quickly, Galileo was able not just to reproduce the telescope but to improve it; his own version achieved at least 8x magnification. He presented his device to the Doge (the highest official in Venice) and secured a truly impressive salary for life from the Venetian state. Mere weeks later he received word from the court of the Medici family, back in Galileo's home of Tuscany, that they wanted a telescope of their own. The Venetian leaders, however, had ordered Galileo to keep his improved telescope a secret, to be manufactured only for Venetian use, and Galileo obliged, at least temporarily.

In 1610, when Galileo turned his improved telescope on the heavens, he realized almost immediately that some long-standing beliefs about the nature of the universe were incorrect. For centuries, philosophers had followed Aristotle in assuming that the planets were perfect spheres moving in perfect circles. When he trained his telescope on the Moon, however, Galileo quickly realized that he was not looking at a smooth, perfect surface. Instead he saw mountains and craters and huge plains, a landscape that looked much more like the surface of the Earth than the featureless sphere described by Aristotelian cosmology. Galileo's hand-drawn illustrations of those first glimpses of the Moon are some of the most compelling and beautiful images in the history of ideas. Looking at them now, we can imagine his awe and excitement at being perhaps the first person in human history to see the surface of a celestial object with any detail (Figure 4.7).

Around the same time he also turned his telescope on Jupiter and noticed, over a period of several nights, that a group of four stars were always visible close to the planet. Their positions relative to Jupiter changed every night – sometimes one was to the left and three to the right, sometimes it was two and two or all four on one side – and it did not take long for Galileo to guess that he was watching four smaller bodies orbiting Jupiter. These were moons, circling around Jupiter just like our solitary Moon orbits the Earth. In fact, what Galileo observed were the four largest of Jupiter's satellites (later named Io, Ganymede, Callisto, and Europa), still known by astronomers today as the Galilean moons.

This single discovery is rightly known as one of the most important moments in our collective understanding of the universe. For the first time, there was empirical proof that the Earth was not at the center of all celestial motion. Those four tiny stars moving around Jupiter not only called into question the ancient structure of the cosmos described by Aristotle, Ptolemy, and countless others, but also raised the intriguing possibility that the Earth might move as well.

As soon as he realized the implications of his discovery, Galileo knew that he had to rush his discoveries into print. He was not the only person with a telescope, after all, and others already could have made the same observations. As quickly as possible, Galileo published

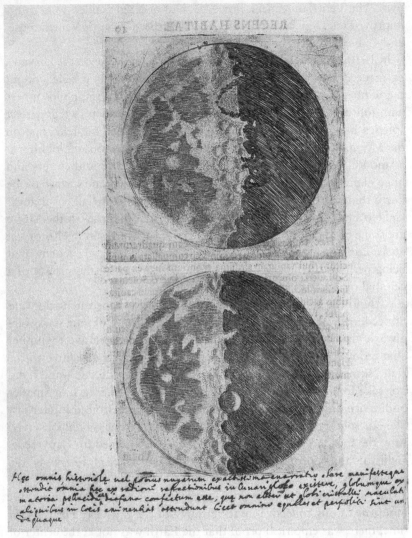

Figure 4.7 Galileo's sketches of the surface of the Moon as observed through his telescope. From his *Sidereus nuncius*, 1610.
Photo by DeAgostini/Getty Images

his *Sidereus nuncius* or "The Starry Messenger," which appeared in 1610 and almost immediately became a huge sensation across Europe. It catapulted Galileo to instant fame and secured him a position as one of the most important thinkers of the century. His interest in publishing his observations was not simply about sharing his exciting discoveries with others, however; he also saw this as an opportunity to attract the interest of a powerful patron, specifically the Grand Duke of Tuscany, Cosimo II de' Medici (1590–1621). Galileo had tried to win the support of the Medici family before, with limited success, but he suspected this would be different. Before he published his discoveries, Galileo wrote to the Grand Duke and offered to name the four stars orbiting Jupiter after Cosimo and his three brothers, a truly spectacular piece of flattery. His offer was met with swift and enthusiastic agreement, and when the *Sidereus nuncius* was published it named the four moons of Jupiter "the Medicean stars." In return, Cosimo appointed Galileo as mathematician and natural philosopher to the Medici court. The title of "natural philosopher" was especially important to Galileo because it represented a promotion of sorts. He was no longer a mere mathematician.

There was an important difference between how Galileo's claims were viewed by his contemporaries and how those of Copernicus had been received some seventy years earlier. Copernicus's argument had been accepted by others only very slowly, and in fact Galileo had been one of a relatively small number of people who believed in the reality of a heliocentric cosmos when he started his academic career toward the end of the sixteenth century. Galileo's claims, however, were different. While many people initially questioned them, they were based on observations that could be corroborated by anyone with a sufficiently powerful telescope. Copernicus had already done the heavy lifting, theoretically speaking. It now fell to Galileo and others to provide empirical evidence for heliocentrism.

The problem was that the moons of Jupiter, while important, did not prove the existence of a heliocentric cosmos. Galileo kept searching until he found something that did: the phases of Venus. You have probably seen Venus before – from the Earth it usually looks like a very bright star – but it is only through a telescope that you can see that it exhibits phases like the Moon. Those phases are important because their existence had been predicted by Copernicus's heliocentric model of the cosmos. If that model was correct, Venus should exhibit the full range of phases, progressing from "new" to "full" and passing through two different crescent phases as it moved from a position between the Earth and the Sun to a position on the far side of the Sun. In Ptolemy's geocentric model, however, Venus only exhibited partial phases because its orbit was always on the near side of the Sun when viewed from Earth. When Galileo turned his telescope on Venus and saw that it exhibited the full range of phases, he knew that he had found something very important.

With the support and protection of the Medici, Galileo quickly became a sensation. He traveled to Rome to demonstrate his telescope to audiences that included cardinals and other members of the Catholic Church, and he enjoyed the kind of fame and recognition that most astronomers could only imagine. Not everyone was convinced by Galileo's discoveries, however, and some were openly concerned about what they might mean. In 1613, one of Galileo's former students, the mathematician Benedetto Castelli (1578–1643), wrote to his mentor after attending a dinner hosted by Cosimo II de' Medici at which the Copernican theory was discussed and vigorously attacked by a visiting professor of philosophy. After dinner, the Grand Duchess Christina, mother to Cosimo, had summoned Castelli and asked that he explain why the idea of a moving Earth did not conflict with Scripture. She was understandably concerned about the reputation of her family, as Galileo was a prominent client of the Medici and had argued publicly for the reality of the Copernican system, a position that could put them all in direct opposition to the Catholic Church. As Castelli later wrote to Galileo, he convinced almost everyone present that the two sides could be reconciled but the Grand Duchess still had doubts. Castelli urged Galileo to reassure Christina that there was nothing impious or problematic about his ideas.

The result was Galileo's impassioned *Letter to the Grand Duchess Christina*, a short essay that attempted to reconcile the Copernican theory with Christian scripture. Galileo wrote it in 1615, and though he addressed it to Christina it also circulated publicly, a practice that was not uncommon at the time. Galileo knew that this was his chance to confront his many critics and demonstrate that a defense of Copernicanism was not necessarily an attack on the Christian scriptures or the declarations of the Catholic Church. The tone of the *Letter* was combative rather than conciliatory, however. Galileo felt that he and his work had been maligned by those ignorant of both astronomy and theology, and he said so very clearly.

The *Letter* attempted to do several things at the same time. First, it established Galileo's belief that the Copernican system was physically real and not an abstract or theoretical concept. In writing this, Galileo all but guaranteed that his claims would conflict with the official doctrine of the Catholic Church. Second, he used the *Letter* to suggest that his opponents – those who tried to, in his words, "damage" and "destroy" him – had failed to interpret Scripture correctly. Galileo believed that the Bible can never be untrue, but that its stories and parables had been "accommodated" to suit the limited knowledge of an earlier time. It spoke of the Earth remaining stationary because that is what people believed when the Bible was first recorded, and in order to make its lessons properly understood the Holy Spirit – which had inspired the authors of the scriptures – had adjusted its language to accommodate the beliefs and knowledge of humanity. It was only after Galileo made his observations that people had proof that the Earth moves. According to him we must read the Bible not as a document that literally and faithfully describes the physical nature of the universe, but as a series of lessons designed to bring us closer to God and to ensure our salvation. As many English translations of the *Letter* have phrased it, "The intention of the Holy Ghost is to teach us how one goes to heaven, not how the heavens go."

What Galileo proposed in the *Letter* to Christina was a separation of natural philosophy and theology. Each had different purposes and methods, and one should not be used to critique the other. The Bible was not a philosophical text, and to use it as such was to misunderstand its real purpose. He also argued that astronomers and philosophers should not be forced to treat their arguments and observations as hypothetical theories or, in his words, "mere fallacies." His main goal, however, was to establish the authority of philosophers to comment on and explain the structure of the cosmos while challenging the authority of theologians to do the same thing. In a sense, Galileo suggested that theologians stick to their own business and let the philosophers go about explaining the world.

In his *Letter to the Grand Duchess*, Galileo could not have produced a more problematic argument, at least from the perspective of the Catholic Church. Its appearance in 1615 marks the beginning of Galileo's real troubles. It was seen by some influential members of the Church as an attack on their authority and, more seriously, on the authority of Scripture, the revealed word of God. This is when the Inquisition first took real notice of Galileo's ideas. In 1616, the Inquisition declared that heliocentrism was a formal heresy because it contradicted the obvious meaning of Scripture. This was a serious development because it would implicate anyone who supported heliocentrism as a potential heretic. At the same time, Pope Paul V (1550–1621) asked Cardinal Robert Bellarmine (1542–1621) to meet with Galileo and make sure that he understood the consequences of supporting the physical reality of heliocentrism.

After summoning Galileo to Rome, Bellarmine had one instruction for him: that he "not hold, teach, or defend" heliocentrism in any way, "either orally or in writing." He was permitted to discuss it as a theoretical possibility, but forbidden to treat it as a physical reality. Galileo agreed – he had no real choice – and carefully avoided the subject for several years afterward. This changed when, in 1623, Maffeo Barberini (1568–1644) became Pope Urban VIII. Barberini had been an admirer of Galileo's for years, and he was considered a highly educated man with a particular interest in natural philosophy. When he learned that Galileo had started writing a new book laying out his ideas about the structure of the universe, Urban asked him to present arguments both for and against heliocentrism while being careful to avoid supporting it as a physical reality. He also wanted Galileo to include Urban's own ideas on the subject, to ensure that the book gave equal weight to both sides.

Thus began Galileo's final and most famous encounter with the Catholic Church. In 1632 he published his *Dialogue Concerning the Two Chief World Systems*, usually referred to today as the *Dialogo* or the *Dialogue*. Importantly, he wrote this work in Italian rather than in Latin, making it accessible to a wider audience at the time. This is the text that set Galileo on a collision course with the Roman Inquisition and the pope himself, and that cemented his reputation in later centuries as a bold advocate for science in the face of religious intolerance and persecution.

The *Dialogue* was written as a conversation between three fictional individuals debating the two main cosmological systems accepted at the time, Ptolemaic and Copernican. The philosopher defending Copernicanism was called Salviati; his adversary was Simplicio, another philosopher who supported the ideas of Ptolemy and Aristotle. Between them was the figure of Sagredo, who was not a philosopher and who represented a neutral position, at least at the beginning. Over the course of several days, both Salviati and Simplicio presented their arguments in favor of their respective systems, with Sagredo asking questions and generally trying to understand what each cosmology involved. There was (and still is) some controversy about the character of Simplicio; more than a few people suspected that his name was an allusion to the Italian word "semplice," which means "simple" or, more properly, "simple-minded." That Galileo might have chosen a derogatory name for the character defending the Ptolemaic system was problematic enough, but still more problematic was the fact that Simplicio repeated ideas in the *Dialogue* held not just by the Catholic Church but, more specifically, by Urban himself.

At the conclusion of the *Dialogue*, the character of Sagredo is persuaded by the Copernican arguments of Salviati, a fact that surprised no one. It seems clear that the *Dialogue* was intended by Galileo to be a fairly one-sided defense of the Copernican system, with Simplicio's arguments dismissed with ease and Salviati's own arguments given preeminence overall. Worse, at least from the perspective of the Church, was that Galileo had conveyed the ideas of Urban VIII, as the pope had requested, but he had put them in the mouth of the foolish Simplicio. Urban had been an ally of Galileo's, but he saw the *Dialogue* as both a public declaration of support for heliocentrism and a personal insult. As a result, he refused to protect his former friend from the inevitable consequences of his work.

Reaction to the *Dialogue* was swift. Galileo was summoned to Rome and, in 1633, placed on trial for heresy. The outcome of the trial was a foregone conclusion. It was clear that Galileo had violated the prohibition against defending the Copernican theory handed down first in 1616 by Cardinal Bellarmine and then reinforced in 1624 by Urban VIII. By Galileo's own account, he was surprised and alarmed by the Church's vehement reaction; he had probably counted on Urban's political support to protect him from serious repercussions, and when that support vanished he found himself almost entirely alone. He insisted throughout his trial that he did not believe in the physical reality of the heliocentric cosmos and that, since 1616, he had never supported the Copernican theory. He did eventually admit that someone *might* be persuaded to believe that the Earth moved after reading the *Dialogue* but that this had not been his goal in writing the book, and he continued to repeat this even when threatened by torture.

Ultimately, Galileo was found "vehemently suspect of heresy," which marked his crime as far more serious than typical, run-of-the-mill heresy. He was forced to deny publicly that the Earth moved, and was then placed under house arrest for the rest of his life. The *Dialogue* was banned in territories controlled by the Catholic Church and publication of any future works by Galileo was forbidden. He lived for another ten years after his trial, locked away in his house outside the city of Florence where he continued to perform experiments in natural philosophy. In 1638 he published his last work, the *Discourses and Mathematical Demonstrations Relating to Two New Sciences*, usually known today as the *Two New Sciences*, an important study of physics. While he had been forbidden by the Roman Inquisition to publish any other works, Galileo was able to secure the help of a publisher in Holland, which was far enough from Rome to make the Inquisition's censure relatively easy to ignore.

Science versus Religion?

The story presented in this chapter is a complicated one. It involves both science and religion, but it is not *about* science and religion – or, to be more specific, about a clash between the two. This was not a fight between winners and losers, or between "right" and "wrong." Instead, this is a story about power, tradition, and authority, about who gets to decide what is true and on what grounds. The cosmos described by Galileo looks very similar to the solar system with which we are familiar today, and in the intervening centuries we have accumulated huge amounts of information that have only deepened our collective understanding of the universe. But even with these advances in knowledge, the deeper issues that motivated the encounter between Galileo and the Catholic Church remain largely unresolved in our modern world. Whether in politics, economics, religion, or science, one can find competing claims to authority and "truth," many of them contradictory.

If this is a story about power and, by extension, the ability to proclaim the nature of reality, then we can look at the different perspectives that existed in premodern Europe and understand them on their own terms. Organized religion, exemplified here by the Catholic Church, had an interest in preserving the *status quo* for many reasons, some of which were undeniably self-serving. As an institution steeped in centuries of tradition, the Church had a significant investment in promoting and upholding traditional philosophies of nature and their descriptions of the universe. The ideas of Aristotle and Ptolemy were still taught in virtually every European university well into the seventeenth century, making the Church's allegiance to these ideas understandable. At the same time, the Church also recognized another source of authority, the Christian scriptures, which stated clearly that the Earth did not move. On both philosophical and theological grounds, then, the Church's position on the immobility of the Earth was reasonable by the standards of the time.

The danger, at least as Galileo saw it, was that the Church presumed to dictate how other people could study the universe. Its popes and priests had good, well-founded reasons for seeing the universe as they did, but Galileo believed that it was not their place to enforce natural philosophical claims about the cosmos, nor to condemn ideas that differed from their own. What Galileo did, ultimately, was transform this from a debate about the nature of the heavens into a much deeper and more important debate about power and authority in seventeenth-century Europe. He believed that the Christian scriptures were not concerned with natural philosophy. Likewise, he argued that the purpose of the Church was to interpret Scripture and help people apply its teachings to their daily lives, not to study the physical universe. In other words, Galileo argued that the Church lacked the authority to make serious and believable claims about the structure of the world.

In Galileo's insistence on drawing a firm line between theology and natural philosophy, one might see the beginnings of the modern, secular belief that science and religion should be understood as fundamentally different endeavors. But of course, Galileo's arguments were self-serving in their own way. He wanted the freedom to say what he wished and to work without interference from the Church. Again, while we might sympathize with him, it is important that we understand how truly radical his position was for the time. Daily observations of the skies appeared to confirm, over and over again, that the Earth was immobile. For many Europeans, the idea that one man peering through a telescope could contradict the collective experiences of millions of people across thousands of years seemed absurd. Imagine if you were told that a single piece of technology, used by just one person, proved that everything you know about the world is wrong. Most of us, I think, would behave just like many people did in the seventeenth century.

The discovery of a new cosmos marked a moment of profound importance for premodern Europe, but also highlights for us today how fragile and tenuous claims to "truth" can be. That fragility remains on display in our own modern debates about the relationship between science and religion. There is little doubt that Galileo would be as outraged by attempts in recent years to legislate the teaching of creation science in classrooms as he was by the attempts by the Catholic Church to police the practice of natural philosophy in the seventeenth century. Understanding the fraught relationship between religion and science in the modern world requires more than an appreciation of their differences. Instead, we must place their relationship in the wider contexts of authority, power, and tradition, as we have for Galileo.

5 | Looking for God in the Cosmic Machine

Following the public disgrace suffered by Galileo at the hands of the Catholic Church, early modern Europeans were understandably concerned about any philosophical ideas that threatened long-standing assumptions about God or the Church itself. Those concerns were only heightened by a new kind of natural philosophy that emerged around the same time as Galileo's trial, known by historians today as the *mechanical philosophies* of nature. They changed how people understood and studied the universe, but also called into question whether God still had a role to play in the world these philosophies described. How different people tried to reconcile these ideas, and their wider implications for the relationship between religion and natural philosophy, will be the focus of this chapter.

At their simplest, the mechanical philosophies sought to explain every natural phenomenon as the result of two things: *matter* and *motion*. Everything in the world, from the orbits of the planets to the bodily workings of a flea, could be understood as the movement of tiny pieces of matter. It was a powerful and ambitious idea because it provided a universal means of understanding nature, a way of taking an enormous range of different phenomena and giving them a single, elegant explanation. While the mechanical philosophies proposed compelling explanations for natural phenomena, however, they also raised a significant problem for early modern people. If everything in the universe can be understood as the interactions between moving pieces of matter – if, in fact, the universe is one vast and intricate machine – then what place is there for God? By reducing all phenomena to the motion of matter, mechanical philosophers inadvertently made room for the possibility that God was no longer a necessary presence in the universe.

This chapter explores the fundamental problems raised by the mechanical philosophies before examining the work of two different people: Pierre Gassendi (1592–1655) and René Descartes (1596–1650). They were contemporaries who each developed a new kind of natural philosophy that explained a wide range of phenomena as the motion of matter. Their systems were very different from one another, but together they had a significant influence on the ideas of later thinkers. They also shared an important similarity: both Gassendi and Descartes were deeply concerned about proving the presence of God in their mechanical systems. Again, the ways in which they addressed that concern were different, but together they illustrate how religious and theological questions became central to these new philosophies of nature.

The (Holy) Ghost in the Machine

There is no doubt that the mechanical philosophies had the potential to provide useful and powerful ways of understanding the universe, especially when contrasted with the most widespread system of natural philosophy that preceded them: namely, Aristotelianism. One reason Aristotelian philosophy survived as long as it did was because someone could use it to explain a wide range of different phenomena. An Aristotelian could explain why heavy objects fall to the ground, why the heavens move, why air is lighter than water, and why dogs give birth to other dogs and not cats or snakes or birds. Its flexibility is why Aristotelianism worked for almost 2,000 years as an explanatory system. Aristotelianism also had some fundamental problems, however. For example, the physics that described objects and motions on Earth was different from the physics that operated in the heavens; explanations that worked in one realm rarely did so in the other.

The mechanical philosophies overcame this and other problems by proposing explanations that truly were universal. The processes or laws that governed an oak tree's growth from an acorn were the same processes that dictated how the planets moved or how the human body worked. Because everything in the physical universe was made of the same kind of thing (tiny pieces of matter) that moved according to a common set of rules or principles, it became possible to relate different phenomena or processes to one another within a single system. This was revolutionary in itself. Of course, no system is perfect, and mechanical philosophers struggled to explain some particularly difficult phenomena like magnetism or gravity, both of which were challenging to conceptualize in purely material terms. Nevertheless, these attempts to establish universal explanations for nature marked an important development in European intellectual history.

For all of the explanatory power offered by these philosophies, however, there remained a fundamental problem: How does God fit into a mechanical system? Finding a role for God at the beginning of the universe is easy – someone could assume that He created matter and first set it into motion. One can also assume that He established certain rules or laws that govern how matter moves on its own, because the alternative is that God Himself has to keep moving every piece of matter all the time and no one thought that was likely. Once everything is moving according to those laws, however, God's intervention is no longer necessary. Being perfect, He would not have created an imperfect system that required tweaks or maintenance, so matter will continue to move on its own without outside intervention forever. And if there is a grand design or purpose that dictates how matter moves and what it forms or becomes across spans of time, presumably God – having perfect knowledge of the future – would have created the universe with that purpose already built in, making His guidance also unnecessary. Thus, one is left with a complex, beautifully elegant machine capable of running on its own for the rest of time, and a Creator with no discernible purpose once everything starts to move.

Particular events, such as miracles, could provide evidence that God was present and intervening in the world because miracles violate the natural order and so must be the work of God, who is the only being capable of violating that order. If someone accepts that miracles occasionally happen, then they would have proof that God is still out there and intervening every once in a while. But what if someone believes that miracles have ceased? This was one major point of contention between Catholics and Protestants in early modern Europe. While Catholics believed that miracles continued to occur, most Protestants believed that miracles had ceased following Christ's ascension into heaven and would not resume until his return to Earth. This meant that, for many Protestants, miracles *could not* be used as proof that God was still present. As the mechanical philosophies became more influential, their adoption by increasing numbers of people brought many of these questions and issues to the forefront of European intellectual culture. This moment, when educated Europeans first began to question the place of God in the wider universe, gave rise to some of the profound changes wrought by the Enlightenment in the eighteenth century.

Pierre Gassendi and René Descartes were not the only people who proposed mechanical explanations for natural phenomena in the seventeenth century, but their philosophies were the most influential. Both systems were "mechanical" in that both tried to explain everything in nature in terms of the motion of matter, but in fact these two philosophies were very different from one another. Gassendi revived the pagan philosophy of the ancient Greek philosopher Epicurus (341–270 BCE), who had taught that the world was composed of miniscule pieces of matter called *atoms*. Descartes, by contrast, created his own philosophy that conceptualized the universe as a vast conglomeration of tiny bodies that moved according to simple geometrical rules. Despite these differences, however, both Gassendi and Descartes shared two primary goals. The first, which each stated explicitly, was to supplant the philosophy of Aristotle. Like others in the sixteenth and seventeenth centuries, they wanted to find explanations for the world that did not depend upon the philosophies of classical antiquity. Their second goal, and arguably the more important one, was to find a place for God in the cosmic machine.

The Baptism of Epicurus

Pierre Gassendi was a Catholic priest who spent his life in France (Figure 5.1). He might seem an odd choice to resurrect a pagan, pre-Christian philosophy, but he was a humanist and, as such, committed to the revival of classical antiquity. Why Gassendi chose Epicureanism over other ancient philosophies is not entirely clear, though he did believe that its central ideas were particularly compatible with his own Christian worldview. The Epicurean emphasis on pleasure rather than fear, for example, may have seemed easier to reconcile with religion as Gassendi himself understood it. He also appears to have been especially drawn to the Epicurean idea of atomism, which we will explore below. His writings on this point are not entirely clear, however, and scholars have sometimes neglected to ask why Epicurean atomism was particularly appealing. As the historian Margaret Osler once suggested, the idea of atomism is so deeply ingrained in modern physics that some of us forget to wonder why it reappeared at all.

PETRVS GASSENDVS DINIENSIS

Figure 5.1 Portrait of Pierre Gassendi, 1658.
Photo by DeAgostini/Getty Images

What was this ancient pagan philosophy that held such promise for a Catholic priest living in seventeenth-century Europe? Like many philosophical schools in ancient Greece, Epicurus and his followers (known as Epicureans) were interested in how one could live a good life. Each school had a different set of ideas about what constituted a "good life" and how someone could enjoy it, and Epicurus taught that the best and most noble goal of life was the pursuit of pleasure. This was not simple hedonism, however, in which physical or carnal pleasure was the ultimate goal. Epicurus believed that pleasure was most easily attained in the absence of pain and fear, and that what most people feared was death – or, more precisely, what came after death. According to Epicurus, people were so busy worrying about what came after death that they lived unhappy and miserable lives. In order to liberate people from this fear, the Epicureans adopted a strictly *materialist* philosophy. They taught that there was nothing in the world except the purely physical or material. Even our souls are made of matter. When we die, there is nothing immaterial that survives; there is simply a material body that no longer functions.

For Epicurus and his followers, materialism was the antidote to pain and fear. They believed that everything in the world was composed of atoms, a Greek word that means something like "indivisible." Epicurus believed that these particles could not be broken down further; they represented the most fundamental components of the universe from which everything was made. Thus, we need not worry that the matter that composes our bodies will continuously decay and fall apart after we die, an idea that some might find troubling or frightening. Instead, when we die the atoms that compose our bodies and souls disperse to become new objects. The torments of the afterlife should not concern us because there is nothing left of us to torment, and thus no afterlife waiting for us. Because there is no afterlife in which souls linger after death, the gods – wherever they are – do not punish the dead for transgressions committed while they were alive. Likewise, we need not worry that our parents or children or friends are suffering those torments themselves. Instead, we can relax and know that after death we will simply dissolve into everything around us.

It is not hard to see why some people might find this a comforting philosophy, and in fact Epicureanism became quite popular, surviving for centuries. The poet Horace (65–8 BCE), who gave us the memorable aphorism *carpe diem* or "seize the day," was influenced by Epicurean philosophy, and the Roman poet Lucretius (d. 55 BCE) summarized Epicureanism in his *De rerum natura* or "On the Nature of Things," which survived into the early modern period and inspired the work of Pierre Gassendi in the seventeenth century.

According to Epicurus, atoms move continuously through what he called the void, which was literally nothing. The void is everywhere, though to us the world seems to be filled with real and tangible objects and not empty space because we perceive only objects created by atoms and not the nothingness through which they move. Atoms themselves are indivisible, but they can be different sizes and shapes and weights, and as they move through the void they come into contact with one another in different ways. Their collisions and interactions are what form the basis of reality as we know it.

Epicurus, however, was worried about the problem of determinism. If everything is the result of the movement of atoms and atoms can only move in particular ways, then everything is essentially predetermined. For example, if atoms can only move in straight lines then the chance of their colliding with one another is limited, which means that reality itself is also limited. Only certain kinds of substances will exist, and they will behave only in certain ways. Epicurus was especially concerned about what this meant for the human mind, which was also a material thing that responded to the movement of atoms. If that movement was prescribed or limited by the physical laws that governed how atoms interact, then the mind was not actually free. At the very least, there were only a finite number of possibilities or choices available to the individual, something that Epicurus absolutely rejected. As a result, he introduced the idea of the "swerve," or *parenklisis* in the original Greek. This was a random event experienced by atoms in which they mysteriously altered their trajectory now and then, swerving to create new paths and new possibilities. This was how Epicurus tried to account for randomness and accidents in nature, but also how he sought to guarantee the existence of free will.

Epicureanism might have disappeared into obscurity like other ancient schools of philosophy, but many of its fundamental teachings were recorded by the Roman poet Lucretius. This was important because Lucretius, living and writing in the Roman Empire, used Latin to explain Epicurean atomism in his *De rerum natura*. While Greek literacy faded in Europe following the collapse of the Western Roman Empire, Latin remained in use among the educated and the literate, and this is probably why someone, or a series of someones, bothered to copy *De rerum natura* through the centuries and why it survived, though unknown and unread, for almost 1,500 years. A copy was finally discovered in 1417 by an Italian scholar named Poggio Bracciolini (1380–1459), sitting in a German monastery where it is likely no one had looked at it in centuries. That single, lucky discovery returned the ideas of Epicurus to Europe, and some 200 years after its recovery the writings of Lucretius would inspire Pierre Gassendi to imagine a world composed of atoms.

Historians of science have sometimes described Gassendi as "baptizing" the pagan philosophy of Epicurus – making it Christian, in other words. Epicurus first proposed his ideas several hundred years before Christianity existed and, while his philosophy made sense in the pagan culture of the Hellenistic world, it required serious effort to make it appropriate for the Christian Europe of the seventeenth century. Fortunately, Gassendi was equal to the task. There were many theological problems with the Epicurean philosophy, at least from Gassendi's Christian perspective. For example, Epicurus had taught that there were many gods, that the universe and its atoms had always existed and always would exist, and that the physical world was the result of chance rather than divine providence or design. We have also seen that Epicurus denied the immateriality and immortality of the soul. These were not small problems, and even now some historians wonder whether Gassendi truly baptized Epicurus or, as Maggie Osler has put it, drowned him in the baptismal font. This is because Gassendi was obliged to make some very big changes to Epicurean philosophy, with perhaps the most significant being his repudiation of Epicurus's desire to free people from a fear of the gods or the afterlife. As a Catholic priest, Gassendi could not deny the existence of an afterlife or the possibility of eternal damnation for sinners.

Most of what was left after Epicurus's "baptism" was natural philosophical, primarily his claim that the physical universe could be reduced to physical or material atoms moving through a void. While Epicurus had taught that the motion of atoms was the first and only cause for natural phenomena, however, Gassendi modified this idea to align it more properly with Christian theology. In Gassendi's philosophy the first or primary cause was God, and all natural phenomena then had as secondary causes atoms moving through the void. The task of the natural philosopher was to study and understand those secondary causes. Through studying those causes, as well as noting the presence of order and design evident in nature, the philosopher would also find confirmation for the presence of God. Gassendi rejected the Epicurean idea that the complexity and order present in the world could be the result of mere chance; on the contrary, he insisted that the world's innate order was inescapable proof for the existence of a divine Creator.

This argument suggests that Gassendi was an empiricist. He believed that the properties of atoms such as size or heaviness could be established not merely by reasoning or thinking about them but by active observation and investigation. Likewise, proof for God's existence was attainable, at least in part, through the observation of nature. Gassendi was far from the first person to suggest that studying the natural world was one means of knowing God; medieval theologians had often argued the same thing to justify their pursuit of natural philosophy. Like Gassendi, René Descartes also believed that a person could establish the properties of matter through observation, but he also claimed that one could understand the nature and properties of matter through nothing but the application of reason and theoretical mathematics.

Figure 5.2 Portrait of René Descartes, c. 1630.
Photo by Hulton Archive/Getty Images

The Cartesian Cosmos

Descartes, living in Catholic France, was affected deeply by the Church's treatment of Galileo, and particularly by his trial (Figure 5.2). Descartes had started preparing his description of the cosmos in 1629, in a work called *Traité du monde et de la lumière* or "A Treatise on the World and on Light," usually known today as *Le Monde*. The trial of Galileo in 1632 caused him to delay its publication, however, because *Le Monde* argued that the cosmos was heliocentric. Descartes feared that his work would be censured by the Catholic Church and that he himself might face accusations of heresy. In fact, he was so worried about the influence of the Church that he left France and moved to Sweden, a Protestant country, where he remained for the rest of his life. Portions of *Le Monde* were published in the 1640s, with the remainder finally appearing in print in 1664. A complete edition of the work was published only in 1677, almost thirty years after Descartes's death.

According to Cartesian philosophy the physical universe is composed entirely of small pieces of matter. Many early modern people called these *corpuscles* and understood them as being distinct from the atoms of Gassendi. Whereas an atom is indivisible and cannot be made smaller, the term "corpuscle" simply referred to a tiny body, and according to Descartes such bodies could be divided repeatedly under certain conditions. He believed that corpuscles possessed a limited range of physical properties such as size and shape, all of which could be expressed mathematically. He also taught that the essence or defining quality of all matter is that it is extended in space. Any object that partakes of matter must also possess extension along three dimensions, and like almost everything in Cartesian physics, extension could be measured and expressed mathematically. Descartes described objects that exhibit extension as *res extensa* – basically, "extended stuff." The whole of the physical universe is composed of *res extensa*, which behaves according to specific laws and principles. The only other substance in the universe is *res cogitans*, "thinking stuff," which we will examine shortly.

Both Descartes and Gassendi described natural phenomena as the result of matter moving, but while Gassendi believed that his atoms moved through a void, Descartes argued that there was no such thing as empty space. In fact, the universe understood by Descartes was literally full of matter; it is what philosophers call a *plenum*, derived from the Latin word meaning "full." He imagined that corpuscles were extended bodies of different sizes, with smaller corpuscles occupying the spaces left by larger ones until no empty spaces existed between them. Because everything was completely full of matter, Descartes believed that the primary kind of motion in the universe had to be circular, an idea that made sense geometrically. Since there is no empty space, corpuscles can only move into space vacated by another corpuscle, creating a situation in which all movement becomes circular as each displaced corpuscle follows the motion of another. This gave rise to the vortex theory of planetary motion in which Descartes described the solar system as a series of bands or currents of circular motion that carry the planets around the Sun. This was one of the most influential aspects of Cartesian physics and had a strong influence on the theories of later thinkers (Figure 5.3).

While Descartes imagined that the universe was completely full, he also believed that matter moved in ways that one could understand and predict using mathematics, and more specifically, geometry. In fact, Descartes argued that corpuscles behaved according to three basic laws or principles:

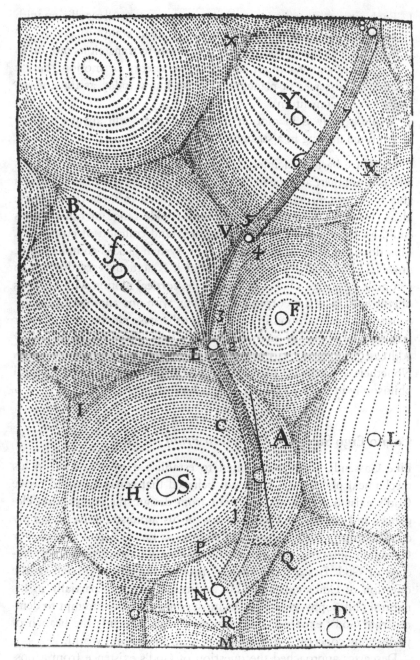

Figure 5.3 The Cartesian universe, with matter moving in vortices around different stars. A comet traces a path starting at "N." From Descartes's *Epistolae*, 1668.
Photo by Oxford Science Archive/Print Collector/Getty Images

1. Each piece of matter persists in its current state until a collision with another piece of matter causes it to change its state.
2. A collision between bodies forces one body to transmit its motion to the other, and it cannot impart more energy or motion than it already possesses (a principle that we know today as the conservation of momentum).
3. While a body is in motion its various parts will always tend to continue moving in a straight line.

These are basic principles, easily imagined by rolling billiard balls into one another, and the genius of the Cartesian system is that it was able to reduce virtually all natural phenomena to these three simple ideas. Heat, light, sound, gravity, and a wide range of other things could all be explained by the movement and collision of corpuscles. This does not mean that the Cartesian system lacked complications, however. Philosophers have noted that it is actually very difficult to reconcile Descartes's belief in circular or vortical motion with his laws of collision. How are corpuscles colliding with one another if they are constantly swept around in vortices or otherwise moving in a circular fashion? It is one of many questions that Descartes never really addressed and that has puzzled philosophers for hundreds of years. Nonetheless, Cartesian physics had a significant influence on natural philosophy in the seventeenth and eighteenth centuries.

Cogito Ergo Sum: *The Cartesian Proof for God*

Like Gassendi, Descartes wanted to establish a role for God in his mechanical philosophy. Gassendi believed that someone could prove God's existence with an argument from design; the complexity, elegance, and order that Gassendi saw in his atomistic universe simply could not have been the result of chance, as Epicurus had argued. For Gassendi, observing nature was the best way to apprehend the presence of God. Descartes, however, took a different approach to the question of God's existence. He addressed this only partially in his natural philosophy, reserving his most comprehensive attempt to demonstrate God's existence for his *metaphysics*, which is the branch of philosophy concerned with things beyond the natural world.

Descartes approached the question of God's existence from a position of extreme skepticism. Before this, most philosophers had believed that our senses and minds are capable of perceiving reality as it truly is – in other words, that what we perceive is real and true. During the Renaissance, however, humanist scholars recovered works written by ancient followers of what is called philosophical skepticism. The Skeptics (who, like the Epicureans, had founded their philosophical school in ancient Greece) had taught that someone could attain a good and happy life if they accepted that they knew nothing, or at least that they knew nothing for certain. By embracing this fact, the Skeptics argued that we would no longer waste our time pursuing "truth" and could instead remain ambivalent about the nature of reality. This was an extreme philosophy in many ways, but by the seventeenth century a more modest form of skepticism had become popular among the educated classes. More and more philosophers began to accept that both our senses and our minds are fallible, and thus cannot necessarily perceive what is real and true. This was a troubling idea for Descartes. If both our senses and minds are unreliable, how can we know anything at all? The answer he devised remains one of the most important philosophical exercises in the seventeenth century, and also demonstrated, at least for Descartes, that God truly exists.

Being a committed skeptic, Descartes examined and discarded every statement of fact as potentially flawed and uncertain, but in doing so he established one single fact with absolute certainty: He was thinking, and therefore he existed. If everything else was uncertain, at least he knew that there was a mind contemplating these questions and that that mind belonged to René Descartes. This is where we find his best-known declaration, "I think, therefore I am," or in the original Latin, *cogito ergo sum*. This is the start of what Descartes called his *Meditations on First Philosophy*, which carried the subtitle, "In which the existence of God and the immortality of the soul are demonstrated," and which first appeared in 1641. The *Meditations* were an exercise in metaphysics rather than natural philosophy, but their implications for Descartes's materialist philosophy are clear. This was part of his sincere and determined effort to find a place for God in his mechanical universe.

Once Descartes established his own existence as a thinking being, he recognized that whatever was thinking these thoughts was a self-contained entity separate from everything else. Moreover, in thinking about his own thinking, Descartes found that at least some of the thoughts or images in his mind were impressions of objects and phenomena that existed outside of himself. Thus, he established that there were actually *two* things: his mind, and everything else.

Descartes accepted that his mind was limited because there were many things that it did not know or understand. Yet, in spite of these limitations, he found that his mind was capable of imagining a perfect being. He reasoned that a flawed and imperfect mind (which he possessed) was incapable of apprehending perfection on its own; therefore, there must be something perfect that existed *outside* of his mind but which his mind could apprehend. This, in a word, was God. Later in the *Meditations* Descartes strengthens his argument for God's existence by reasoning that perfection implies existence – that is, something that is truly perfect in every way will exist because existence is better (or more perfect) than nonexistence. Therefore, God exists.

The Cartesian *Meditations* are a unique and rather beautiful exercise in applied reasoning, though not necessarily persuasive to everyone. Nevertheless, the *Meditations* are important for a couple of reasons. First, they establish the Cartesian idea that the existence of God and the nature of reality can be proved merely by thinking. Gassendi was an empiricist who looked to the world around him for proof of his ideas, but Descartes believed that all one needed to prove the truth of his philosophy was to use reason and logic to answer fundamental questions about the universe. He argued that we cannot look for certain proof outside of our minds because our perception of the world is inherently distorted by the weakness and fallibility of our senses. The *Meditations* also elevate proof for the existence of God as crucial to the process of understanding the world, which is no accident. Whereas Gassendi had taken up the pagan philosophy of Epicurus and added the Christian God to it, Descartes devised a philosophy that began with the existence of God and used that one idea to explain and justify everything else.

Later in the *Meditations*, Descartes considered the nature of physical reality. He believed that the best and most perfect way to understand the universe was through mathematics, and specifically geometry. Geometry was abstract and eternal and perfect; its fundamental rules were always true. In embracing this idea Descartes was following in the footsteps of other thinkers of the time, including Galileo (who once claimed that God wrote the Book of Nature in the language of mathematics) and Johannes Kepler (who used geometry to understand and explain the harmonious order of the cosmos). Like them, Descartes believed that principles derived from geometry were, like geometry itself, always true. These included the principles of movement, duration, and extension, all of which a Cartesian philosopher could apply to a world composed entirely of matter.

Thus, the Cartesian universe is composed (mostly) of matter that is extended in space and that behaves according to mathematical principles first devised by God. Those principles, like God, are true and perfect and unchanging. Because matter behaves according to these principles, a philosopher – assuming they understand these principles correctly – can predict and explain the behavior of physical reality. At the same time, the very existence of those principles is itself proof of God's existence, since Descartes believed that those principles flowed directly from the perfect and unchanging nature of God.

Now that we have an idea of how both Gassendi and Descartes envisioned their mechanical worlds, let's take a look at how they and other mechanical philosophers tackled a series of important problems: the nature and existence of the human soul; the origins of life in a mechanical universe; and finding empirical evidence for the supernatural.

The Soul

Both Gassendi and Descartes made space in their respective philosophies for God, and in doing so they had to confront a problem shared by all materialist philosophies – namely, that sometimes they have to account for substances that are *immaterial*. Some materialist philosophies, like that of Epicurus, simply do not accept that immaterial substances exist, which solves that particular problem rather neatly. But most philosophies are not strictly materialist and instead propose that some nonmaterial substances do exist. For both Gassendi and Descartes, living and writing in the seventeenth century, it was crucial that they account for immaterial substances in their mechanical philosophies. We know how they tried to do so in the case of God, and now we will examine their efforts to explain the soul.

The soul is a tricky concept. If you subscribe to a materialist philosophy but also believe that souls exist, then you have to answer a number of difficult questions. For example, if we accept that a person or animal possesses some immaterial component, where is it localized? Is it in one place or is it present throughout that person or animal? If it is present throughout, can we separate pieces from that person or animal that also contain a soul, and does that mean that we now have two souls? If I lose an arm or a leg, is my soul now smaller or otherwise diminished? Does my amputated limb now possess a soul of its own? If the soul is present in just one part of something, what happens if that part is damaged or removed? Is having a soul the same as being alive, or can something live without a soul?

There is almost no end to the questions raised by the idea of the soul, and people have asked many of them at different points throughout history. Aristotle, for example, taught that there were different kinds of souls possessed by different kinds of substances, an idea that was accepted by most medieval and Renaissance philosophers. In fact, he proposed a hierarchy of souls: the vegetable soul, the animal soul, and the rational soul. The vegetable soul is possessed by every living thing and permits the basic functions of life to occur. The animal soul is of a higher order and is possessed, of course, by animals, which are capable of more complex movements and behaviors than plants or worms or insects. Finally, there is the rational soul, possessed only by humans; in fact, for Aristotle the rational soul was the thing that *makes* us human.

In adapting Epicureanism to early modern Christianity, Gassendi faced a particular challenge when it came to the question of the soul. There was no doubt what Epicurus had taught on this subject, which was that soul is mortal and composed entirely of atoms. It did not persist after death, but instead dissolved into its constituent atoms along with the rest of the individual. Obviously, Gassendi would not have agreed with this idea, but because he had rejected most of Aristotelianism he could not draw on the rich tradition of commentaries produced from the Middle Ages onward that reconciled Aristotelian philosophy with Christian theology. Nevertheless, he did get some help from Lucretius, the ancient Roman poet who had translated Epicurus into Latin.

Writing in the last decades before the beginning of the Common Era, Lucretius claimed that the soul was corporeal or material in nature and was composed of two parts: the *anima*, which was the nonthinking part responsible for basic functions of the body, including sensation and nourishment; and the *animus*, which was the rational or thinking part. (Note the gendered distinction that Lucretius used: the rational part of the soul was identified using the masculine noun while the irrational or nonthinking part was feminine.) Gassendi followed this same division and believed, as did Lucretius, that the *anima* was present throughout the body. He argued, however, that the *animus*, or the thinking part, was localized in the head. Unlike Lucretius, Gassendi believed that the *animus* was incorporeal and corresponded most closely to Christian doctrine concerning the human soul. The *anima*, however, could be corporeal since it was shared by nonhuman animals and had no real theological significance.

The *anima* provided vital heat to the animal and allowed it to live; the absence of *anima* was death. The *animus*, however, had to be incorporeal because it was capable of actions that the *anima* was not. Gassendi argued, for example, that the intellect is capable of apprehending itself, which is called *metacognition* in modern philosophy. We are aware that we are thinking, and only an incorporeal or immaterial thing is capable of this kind of action because matter is not capable of apprehending itself, an argument not dissimilar to the Cartesian dictum of *cogito ergo sum*. Gassendi also claimed that, unlike animals (who possess only the *anima*), humans can recognize and think about the existence and nature of universals (for example, the nature of "human-ness" or "flat-ness" or "dog-ness"). Again, because this is a faculty or ability that is not shared by animals, Gassendi reasoned that humans must possess something that is fundamentally different from the corporeal or material *anima*, which is the incorporeal and immaterial *animus*. Moreover, once Gassendi established that humans possessed an incorporeal soul, it was relatively easy for him to argue that this soul is immortal. Because it is incorporeal it is not subject to corruption or dissolution; it has no material parts that can decay or become separated, and so it will remain eternally unchanged.

Importantly, the immaterial nature of the *animus* means that it cannot be studied or understood by Gassendi's mechanical philosophy. As Maggie Osler noted, he willingly diminished the explanatory power of atomism in order to preserve a space for immaterial, nonmechanical substances in the universe. She also observed, however, that even after all of Gassendi's efforts to secure the presence of God and an incorporeal soul, some early modern people remained concerned about the potential for atheism in Gassendi's philosophy. Baptizing a pagan philosophy simply may have presented too many problems in a post-Galileo Europe.

Turning now to Descartes, we find that he differed from both Aristotle and Gassendi on the question of the soul. He believed that

Figure 5.4 The mechanical transmission of sensation from the foot to the brain, resulting in involuntary movement away from heat. From Descartes's *Opera Philosophica*, 1692.
Photo by Oxford Science Archive/Print Collector/Getty Images

the only corporeal beings that possessed a soul were humans, and that animals were little more than machines or *automata* whose movements and behaviors were the result of mechanical principles rather than cognition or even the corporeal *anima* proposed by Gassendi. If an animal tried to avoid pain or moved toward a source of heat, it did so not because of any cognition or thinking on its part but because of a simple mechanical response to stimulus. Human behavior, too, can result from a purely mechanical response – our instinctive recoil from touching a hot object, for example – but, unlike animals, we are also capable of genuine cognition and emotion (Figure 5.4).

You might remember that Descartes divided the universe into two types of "stuff": *res extensa*, or extended stuff, and *res cogitans*, or thinking stuff. This is a philosophical position known today as "mind-body dualism," the idea that mind and matter are two completely different substances, and it is one of Descartes's most important and lasting contributions to modern philosophy. It comes with a built-in problem, however. How do these two different kinds of "stuff" inter-act with one another? It is one thing to claim, as Descartes did, that *res cogitans* exists and is purely immaterial, but it is much more difficult to explain how an immaterial substance can influence a material thing. Descartes never fully resolved this problem, and it remains a point of debate among philosophers today. That *res cogitans* can affect the material world was, for Descartes, a given. God, after all, is an immaterial being who can affect the universe, and there were some Cartesians who speculated that God Himself was composed entirely of *res cogitans*. But because God is largely unknowable and His powers largely incomprehensible to us, His ability to affect the physical universe is not very useful in answering these questions about the relationship between matter and the soul.

The solution that Descartes devised was interesting but also problematic, and it evolved over time. When he first wrote about the soul he did so in his *Treatise on Man*, which he finished sometime before 1637 but which was published in the 1660s, after his death. There, he argued that there was a physical part of the body that housed the soul, because he believed that in order for the two to interact they had to be connected in some way. The place where Descartes located the soul was the *pineal gland*, a small organ buried deep inside the brain. The rest of the body, he wrote, was purely material, effectively a human-shaped machine in which many processes such as breathing, digestion, growth, and reflexes (like flinching away from a hot object) were simply mechanical in nature and did not require thought or the intervention of an animal or vegetative soul. Likewise, impressions received by the senses were propagated along nerves as mechanical impulses (to borrow from modern physics, we might think of them as pressure waves moving through a fluid medium) until they reached the brain, and many of our responses to sense impressions, such as our mouths watering when we smell something delicious, could be explained as purely mechanical reactions to stimuli.

For Aristotelians, all of these bodily behaviors were governed and regulated by the soul, and specifically by either the vegetative or animal soul. Descartes, however, believed that his system was superior because it did not require these different souls to explain basic bodily functions. In fact, the only thing that required a soul was cognition or thinking, which was also the only thing that sets humans apart from other animals. After all, Descartes reasoned, a dog eats and sleeps and grows larger and reacts to stimuli like heat and cold, and it does so in ways almost identical to how our own bodies move and behave. And yet, Descartes believed that we are obviously superior to dogs, and the thing that makes us superior is our mind. Therefore, our bodies might be mechanical in form and function, but our minds must be something else. We no longer share a vegetative or animal soul with dogs; we possess a human soul while dogs have no soul at all.

Following his work in his *Treatise on Man*, Descartes spent a lot of time thinking about the soul. Eventually he suggested that it is present throughout the body in the same way that gravity affects every part of the body. It can do so because the soul is not extended in space, and therefore not subject to the limitations of material or physical objects, but he still believed that it was localized and somehow most "present" in the pineal gland. It was there that some sort of interface between *res cogitans* and *res extensa* existed. Descartes speculated that if corpuscles became small enough they would begin to approach the state of no longer being extended in space, which would bring them closer to the nonextended state of *res cogitans*. But was there a point at which extended matter could actually become "thinking stuff?" According to Cartesian physics, which was very clear that *res extensa* and *res cogitans* were two entirely different substances, the answer was no. This speculation about the pineal gland, while interesting, ultimately did not resolve existing problems involving the interaction between soul and body.

Because Descartes was committed to mind-body dualism, he was not able to provide a satisfactory explanation for the soul's ability to interact with the body. Some of his later followers went so far as to suggest that the only thing that could bridge the divide between the two was God Himself, so that all interactions between the human soul and its body were mediated somehow by God. As a philosophical explanation, however, that is not particularly satisfying. Ultimately, the problem of Cartesian dualism passed into the more general realms of philosophy and psychology where it continues to preoccupy scholars and academics even now.

The Mechanisms of Life

As the mechanical philosophies became more popular, another problem began to emerge: their insistence that matter itself was completely inert. It was essentially "dead stuff," incapable of moving itself. In the elegant mechanical systems envisioned by Gassendi, Descartes, and their followers, this was not necessarily a problem. They argued that

God imparted motion to matter and that basic laws or principles operated to keep that motion going for the rest of time, so matter did not need to be active or move itself. But, some wondered, how can life emerge from the movement of dead pieces of matter? How might the spark of life demonstrably possessed by plants, animals, and people be the result of nothing more than the motion of atoms or corpuscles? It is worth noting that this very question is still an issue in modern philosophy – some scholars continue to ask how the incredible complexity of the human mind can be explained as the electrochemical firing of neurons, the exchange of neurotransmitters, or the movement of particles in the brain. Thus, this was not a problem confined to the seventeenth century; instead, it is an issue that reappears whenever people propose materialist explanations for complex natural phenomena, and particularly for (human) life.

Descartes tried to distinguish inert or dead matter from the stuff that makes humans human; his *res cogitans* was a different kind of thing from the passive and inert *res extensa* that made up all physical objects in the universe. Claiming that there is a different kind of substance – even an immaterial substance – that explains the presence of the human mind or soul, however, is not quite the same thing as explaining life itself using the movement of matter. This led some in the seventeenth century to suggest that matter was not inert and lifeless at all, but possessed some active or "vivifying" principle. This is usually related to the idea of *vitalism*, which holds that material substances possess some vital or animating quality. Vitalism is distinct from, and often proposed as an alternative to, purely materialist explanations for natural phenomena, and examples of early modern philosophers who held views similar to vitalism are Margaret Cavendish (1623–73) and Georg Ernst Stahl (1660–1734).

The idea that matter might be animated in some innate way potentially solved a couple of problems. For one, it provided an alternative to Cartesian dualism in which soul and body were utterly distinct from one another. If matter itself possesses vitalistic or life-like qualities, the line between soul and body becomes less distinct, making explanations for how one affects the other easier to imagine. The notion of vitalistic or animate matter also provides a plausible explanation for how pieces of matter continue to move, and in fact makes it possible to suggest that matter is essentially self-organizing. Mechanical arguments that everything around us is the result of matter bouncing into other pieces of matter seem plausible until we start to notice how incredibly complex the world is. How does something as complicated as the human body arise through the (possibly random) movement of matter, even if it follows mathematical principles or adheres to rigid laws? If we endow matter with some vivifying or animate quality, then we can suggest that it moves and organizes *itself* into the complexity we see everywhere in nature.

Vitalistic matter presents its own problems and challenges, however. Giving corpuscles or atoms some quality similar to life itself could lead to the suggestion that our bodies do not actually require a soul in order to be alive. On the other hand, someone might conclude that souls are composed of this vital matter, turning the soul into a material object, an idea that no good Christian could possibly accept. And inevitably we return to the problem of God's agency. If matter is able to move itself, what need is there for the guidance of a Creator?

The question of how inert matter can explain life's complexity, as well as the argument that matter itself is imbued with some vitalistic or life-like principle, were both responses to the serious existential problems raised by the mechanical philosophies. Both also impinged on religious and theological concerns in the seventeenth century because they concerned the origins and nature of life in a mechanical universe.

Mechanizing the Supernatural

Gassendi and Descartes went to considerable lengths to secure a place for God and the human soul in their mechanical philosophies. Their efforts did not convince everyone, however. If these arguments were not enough, what could they or their followers propose instead as evidence for the existence of God in a mechanical world? The presence of miracles was one possibility; events that overturned natural laws and principles could only be caused by God, since He was the only being capable of overturning those laws, and so miracles could become tangible, empirical evidence for God's presence. European Christians were divided when it came to the question of miracles, however. Catholics believed that they persisted and occurred regularly, but most Protestants did not; they believed that miracles had ceased and would resume only with the Second Coming of Christ.

If miracles were not enough to convince everyone, then perhaps there were other ways to establish the existence of the supernatural realm. After all, while God was the preeminent supernatural being, He was not the *only* being that existed beyond or above nature; there were also angels, demons, and the Devil himself. We saw in Chapter 2 that witches and, more specifically, their sexual interactions with demons may have provided important empirical and tangible evidence that demons really existed, which in turn suggested that other supernatural beings like the Devil and God also existed. A similar line of reasoning became useful to some seventeenth-century philosophers who wanted to retain a place for the supernatural in a mechanical universe.

The problem facing these philosophers was that God, by definition, is difficult to see or understand with any certainty. We might infer His presence from any one of a million phenomena around us, but *proving* that He exists is an entirely different thing. In fact, demonstrating the presence and reality of the supernatural is a problem still faced by virtually every religion in the world today, because that reality does not and cannot rely on conventional types of evidence or methods of demonstration. Belief in the supernatural requires a leap of faith that takes someone beyond the realm of observations and experiments that otherwise allow us to understand the world around us.

But what if there *were* experiments we could perform that provided proof for the supernatural? Experiments and observations became increasingly important to natural philosophers over the course of the seventeenth century, as part of what historians call the "new science." Proponents of this "new science" wanted to devise methodologies that would allow them to study the world in systematic and reliable ways. They were ambitious and motivated enough to try to apply these evolving methodologies to big questions like the existence and presence of God, and this is where our narrative about the mechanical philosophies really gets interesting.

If we put ourselves in the shoes of these early modern philosophers, we find that we have new and powerful ways of understanding the universe but that they raise uncomfortable questions about the presence of God. One means of addressing these questions is to propose amendments to these philosophies that make room for the immaterial soul and work out complicated chains of logic that prove that God exists. Another would be to find observable and measurable events that might also establish the existence of the immaterial and the supernatural. Then we could take these novel methodologies of observation and experiment and apply them directly to the question of the supernatural. If we manage that, we will have a powerful means of combating atheism while also demonstrating the rigor and usefulness of these new methodologies *and* proving that our mechanical philosophies can accommodate an active role for the supernatural.

In fact, one of the more interesting consequences of the "new science" that emerged after the middle of the seventeenth century was the way in which naturalists and philosophers brought its methodologies to bear on phenomena like witchcraft, ghosts, demons, and magic. If empiricism and experimentalism were truly as powerful as their supporters claimed, they should be able to explain even the most mysterious of natural phenomena. On the other hand, if there were phenomena that *could not* be explained by these methods, this might be evidence that some things simply existed beyond the natural world – in other words, that they were supernatural. It was a win-win for supporters of the mechanical philosophies. Even phenomena they could not explain might help them demonstrate the existence of the supernatural, which was a major problem that every mechanical philosopher had to address.

Some of these efforts appear in the work of Joseph Glanvill (1636–80), a member of the Royal Society of London as well as a philosopher and clergyman. Glanvill was not a strict Cartesian but did embrace a generally corpuscularian or mechanical worldview. At the same time, he was fascinated by reports of witchcraft and other possibly supernatural occurrences. In the year after his death, his greatest work on the subject appeared in print, titled *Sadducismus triumphatus*. This was an expanded version of an earlier work, *A Blow at Modern Sadducism*, which he published in 1668. "Sadducism" referred to the beliefs of the Sadducees, an ancient Jewish sect that denied the existence of angels and other supernatural beings. Glanvill was deeply concerned about what he saw as a form of modern Sadducism, which was the skepticism with which people in the seventeenth century approached reports of witchcraft and other, similar phenomena. He believed that skepticism about witchcraft was only a short step from outright denial of the supernatural and, from there, to atheism.

Witchcraft and demonic apparitions were useful to Glanvill for two reasons: they produced tangible, observable effects that the naturalist could examine, study, and record; and they presupposed the existence of the supernatural because both required the active intervention of demons. In his *Sadducismus triumphatus*, Glanvill provided detailed reports of events or instances of witchcraft, hauntings, and demonic possession. Presenting these accounts allowed Glanvill to appeal to new and emerging standards about evidence that were circulating in the later seventeenth century. Thus, *Sadducismus triumphatus* was Glanvill's attempt to persuade other natural philosophers that witchcraft and ghosts were real phenomena worthy of investigation. More importantly, for Glanvill, these phenomena were tangible and observable evidence for the existence of the supernatural in a mechanical universe.

But while demons, ghosts, and witches allowed Glanvill to apply the empirical methods of the "new science" to the question of the supernatural, for others in the seventeenth century this methodology was problematic. The historian Thomas Jobe has shown that a public argument between Glanvill and the chemical physician John Webster (1610–82) about the reality of witchcraft demonstrated how complicated and contentious these issues became when tackled by people with different worldviews. Jobe pointed out the apparent paradox confronted by historians in the twentieth century when they realized that the empirical and careful Glanvill wrote a passionate defense of the existence of witchcraft, while Webster – an advocate for Paracelsian medicine and someone who believed in the weapon salve and other examples of occult activity in nature – denied witchcraft's existence. How could a modern-seeming proponent of the "new science" defend witchcraft while a credulous occultist took a far more skeptical approach?

Jobe framed the debate between Glanvill and Webster as fundamentally concerned with different interpretations of Protestantism, but also pointed out that Webster's skepticism about witchcraft was motivated by his desire to defend the practice of natural magic. Webster believed that magic was not inherently bad or the work of the Devil, but was instead the endeavor described a century earlier by Giambattista della Porta who characterized natural magic as "the practical part of natural Philosophy" and "the survey of the whole course of Nature." In order for Webster to defend magic as natural, he had to sever its associations with demonic intervention, which is why he argued that witchcraft was false.

The encounter between Glanvill and Webster demonstrates how complicated these issues were. In early modern Europe, someone could use empiricism and experiment to defend ideas that might strike us as deeply irrational, like the reality of witchcraft. Likewise, someone could exercise careful and rational skepticism in order to support a belief in occult forces and sympathies. And overlapping with one another, layer after layer, are competing interpretations of religion, natural philosophy, and magic.

We started this chapter with the rise of the mechanical philosophies and the serious problems they raised for philosophers and theologians in the seventeenth century. We have studied different solutions to those problems, some perhaps more compelling than others. It is important to note that, however problematic the idea of a mechanical universe might have been, it did not disappear. More and more philosophers embraced the notion of a world made up of tiny pieces of matter, until we arrive at the universe explained by modern science in which (material) particles interact with (immaterial) forces. The preoccupation with establishing a place for God and the supernatural that we find in the seventeenth century largely disappeared from the study of the universe over the course of the eighteenth century, however, thanks in part to the widening separation between science and religion that characterized the cultural movement known as the Enlightenment.

6 | *Manipulating Nature*
Experiment and Alchemy in the Scientific Revolution

This chapter examines the rise of the "new science" in the seventeenth century and its emphasis on empiricism, experience, and experiment. Many historians of science view this development as the logical culmination of the "Scientific Revolution" that began with Nicolaus Copernicus in the sixteenth century and ended with the wholescale rejection of Aristotelian natural philosophy by the end of the seventeenth century. In fact, that rejection was already underway by the middle of the century. The rise of the mechanical philosophies was an important indication that Aristotelianism was slowly losing its grip on mainstream natural philosophy. The same was true of assertions made by Galileo, Kepler, and others that the Earth moved. In the end, after surviving for more than 2,000 years, the worldview inherited from classical antiquity unraveled surprisingly quickly.

Generally, we can understand the "new science" as a deliberate effort to move away from the philosophy of Aristotle, whose ideas about experience were seen as more and more problematic by naturalists living in the seventeenth century. As early modern people began to embrace new definitions of experience, however, they encountered new problems as well. For example, how could something experienced by just one person be useful to (or trusted by) others? Such problems assumed greater significance in the context of experimentation, which occupied a new and vital role in seventeenth-century natural philosophy. Greater numbers of people began to rely on contrived or deliberate attempts to study how the universe worked, arguing that the manipulation of nature was the best way to understand it. It is an idea that makes intuitive sense to us today, living as we do in a time when experimentation is central to how we understand the practice of science, but it was not without controversy in the seventeenth century.

This chapter also considers the history of alchemy, and particularly its "golden age" in the sixteenth and seventeenth centuries. It is a fascinating subject in itself, but the beliefs and practices of alchemists

161

can help us appreciate how challenging it was to reconcile long-standing traditions of natural philosophical study with the drive to create methodologies that demanded new standards of evidence, practice, and dissemination. The secrecy that had long shrouded alchemy, as well as the mysterious and often unsubstantiated claims of alchemists, eventually came into conflict with the increasingly collaborative and open style of natural philosophy encouraged by individuals and institutions in the latter half of the seventeenth century. This is not a story about a rational and empirical "science" doing away with the irrational superstition of alchemy, however. What actually happened to alchemy was both more complicated and more interesting.

Aristotelian Experience

In order to understand how the notion of "experience" changed in this period, we need to start with the preeminent authority in natural philosophy, and at the beginning of the seventeenth century that was still Aristotle. Experience was central to Aristotelian natural philosophy; Aristotle taught that in order to understand nature, we need to rely on our direct experience of it. He had particular ideas about what constituted an appropriate kind of experience, however. The strongest and most persuasive form of experience was what Aristotle called *universal experiences*, which are multiple experiences that accumulate over time. One simple example is the experience of the Sun rising each morning. We can witness it happening every day, and our shared, collective, and accumulated experiences of the same thing all add up to a universal experience that demonstrates a natural process. Universal experiences allow us to make similarly universal inferences about how nature works. For example, we can say that the Sun always rises in the east and know that this is a fundamental truth about the world as we experience it daily.

Universal experiences were important to Aristotle because they were most likely to capture the way that nature usually or normally works. A phenomenon that happens repeatedly and is experienced by many different people reflects the "ordinary course of nature," in the words of premodern philosophers. Deviations from the norm, especially deviations that only occurred once or twice, were not worth considering. What Aristotle called "accidents" were small deviations that did not affect the essence of something – for example, a black cat and a white cat both share the same essence of "cat-ness" but differ only in their coloration, which is an accident in the Aristotelian sense. Accidents are worth noting but ultimately tell us nothing useful; studying the color of a cat's fur does not help us understand what makes a cat a cat and not a dog or a tulip. Similarly, Aristotle taught that larger deviations from the way nature normally behaves cannot tell us anything useful about nature itself. If we are interested in understanding the world by relying on universal experiences, then this makes sense; by definition, singular or isolated experiences are unable to provide truly universal knowledge, which makes them less persuasive and, ultimately, less useful to the Aristotelian natural philosopher.

Having examined how Aristotle understood "experience," take a moment to think about how experiments work in modern science and medicine. What is the purpose of experiments, and what are scientists actually *doing* when they perform them? We could answer these questions in different ways, but a very simple way of defining an experiment might be something like this: It is a singular experience in which someone establishes natural principles by the careful manipulation of substances and variables. When a scientist conducts an experiment, they have a hypothesis in mind which they then test by seeing how natural processes operate under certain conditions, usually manipulated by the scientist themselves. Experiments form the bedrock of modern science and medicine, and yet, were he still alive, Aristotle would insist that these experiments cannot tell us anything useful about nature. How, then, did we get from Aristotle to the modern scientific experiment?

According to Aristotle, singular or isolated experiences had very little use in determining how nature normally worked, which was the highest standard for knowledge about the natural world. He also made a sharp distinction between *nature* and *artifice*, arguing that natural and artificial objects are fundamentally different from one another. A natural object contains the principles of change and motion within itself; they are an intrinsic part of its being. Objects created by artifice do not contain these intrinsic principles, however. An acorn has the ability to sprout and grow into an oak tree; a chair made from oak does not. For Aristotle, artificial objects copy nature but do not partake of it. This makes them unsuitable for understanding how nature works, because the goal of the natural philosopher is to understand the principles of change and motion inherent in natural objects.

Artificial objects, lacking these principles, are of minimal use in understanding nature.

The Aristotelian problem with artifice is another way in which the practices and methods of modern science differ from early modern natural philosophy. We might make a distinction between "field science," where someone goes out into the world and makes observations of what they find there, and "bench science," where instead someone stays put in a laboratory and contrives experiments to test particular hypotheses or ideas. Of course, science rarely falls into such neat categories, and many scientists use experiments to test hypotheses out in the real world as well as in the laboratory. But it is worth reminding ourselves that for Aristotle and many other premodern people, the simple and passive observation of nature was the gold standard for natural philosophy – it was the best way to understand how nature normally works. According to Aristotle, artificially manipulating variables and circumstances to produce different results (which is how most experimental science works today) was useless because, by definition, doing so distorted nature's workings and forced nature to behave in ways that were not usual or normal. Why would someone want to know how nature *might* behave in this one contrived situation that may never occur spontaneously?

In fact, Aristotle's concerns about the artificial manipulation of nature are worth considering carefully. He was not wrong to question what value there is in encouraging or forcing nature to behave in ways that might never happen outside of those very specific and limited circumstances. If we allow ourselves to think like an Aristotelian for a moment, we might agree that this way of studying the world carries with it some potential problems. Aristotle's ideas about experience and artifice were so convincing, in fact, that they made sense to generations of philosophers across almost 2,000 years. Like him, many of them taught that the artificial manipulation of nature to produce singular or isolated experiences was problematic. On top of this we can add a healthy dose of social snobbery. Working with one's hands was seen as a "vulgar" or lower-class activity, fit for manual laborers but not scholars and philosophers. We caught a glimpse of that attitude in Chapter 3 when we examined premodern medicine and its division between the physician and the barber-surgeon, and it pops up again here because in the early modern period experimentation was perceived by many as too close to manual labor to be entirely appropriate for the refined or educated person.

Figure 6.1 A nineteenth-century engraving of Francis Bacon.
Photo by E+/Getty Images

For these reasons, the deck was stacked against experimentation during much of the history we have covered in this book. Obviously, however, something changed. In fact, a *lot* of things changed, only some of which we can discuss here. The most profound change, simply put, is that by the middle of the seventeenth century more and more people no longer viewed Aristotle as the preeminent philosophical authority that he had been even a few decades earlier. He continued to have his defenders, of course, and the standard university curriculum remained heavily indebted to Aristotelian ideas into the late seventeenth century, but by and large this period witnessed a departure from the philosophy of Aristotle. This created new intellectual spaces in which ideas and methodologies that had received little attention from Aristotelians could now be explored in earnest. The experimental method was one of them.

Francis Bacon and the Inductive Method

When asked about the origins of the modern scientific method, many historians point to the English philosopher Francis Bacon (1561–1626) (Figure 6.1). While he did not single-handedly create the methodology used by modern scientists, he played an important role in its development during the seventeenth century, in part because his ideas influenced some of the first people to apply these methods in a systematic and widespread way. Though Bacon died in 1626 his philosophy continued to gain traction in the following decades, and by the 1650s significant numbers of people were committed to the Baconian program of natural philosophy. One of the earliest "scientific" societies, the Royal Society of London, was founded in 1660 with an explicit and long-standing commitment to Baconian ideals, and experiments were central to how members of the Royal Society both investigated nature and demonstrated their findings to others. It is no surprise, then, that historians see Francis Bacon as the origin of our modern scientific method.

Bacon founded his ideas about experience and experiment on what is known as *inductive reasoning*, or *induction*. Putting it simply, Bacon believed that someone could establish an understanding of the general principles and rules governing the workings of nature by the careful observation of particular objects or events. In other words, if we want to know how nature operates we need to observe many singular phenomena and then add up our observations to produce more general and universal claims. Historians of science often contrast Bacon's method with that of René Descartes, which depended upon *deductive* reasoning in which one starts with general principles and then follows them to particular conclusions. For example, Descartes started with general ideas such as "I am thinking, therefore I exist" and "There is a perfect, eternal being, which is God" and then used those principles to make smaller and more specific claims about particular things in the world. Bacon, however, worked in the opposite direction, beginning with a series of singular observations and then building up to general principles. For this reason, some philosophers think of inductive reasoning as "bottom-up" while deductive reasoning is more "top-down."

In choosing to focus on singular observations, Bacon was of course doing exactly what Aristotle taught his students *not* to do. For Bacon, however, this was a good thing. Like increasing numbers of people in early modern Europe, he wanted to break away from the teachings of Aristotle and other philosophers from classical antiquity and chart a new path for the study of nature. Universal claims were still valuable to Bacon – his ultimate goal was to uncover the general principles that dictated how nature worked – but he believed that the philosopher must start with what was right in front of them: namely, individual or particular things. Like almost everyone else living in this period, however, Bacon was still very much a product of Aristotelianism. It is interesting to note, for example, that both Bacon and Aristotle saw *accumulation* as a key necessity in understanding nature. For Aristotle it was the steady accumulation of shared or collective experiences that gave rise to universal knowledge, while for Bacon it was the accumulation of singular observations. Yet, there are also important differences between these approaches. While Aristotle envisioned natural philosophy as an endeavor that spanned whole generations, building on the experiences and observations of many people, Bacon's method was individualistic; it could be carried out by one person recording and interpreting observations. Really, though, the Baconian method becomes genuinely powerful when it, too, embraces collaboration, and this was the impetus that drove the formation of some of the first scientific societies in Europe.

In order to understand Bacon's kind of inductive reasoning, consider the nature of light. It is a natural phenomenon, of course, but how could someone try to understand it? Using the Baconian method, they would start by making singular observations of light, and preferably of the many variations of light that exist in the world. They would observe sunlight, moonlight, starlight, candlelight, firelight, reflected light, the glowing abdomens of fireflies, and every other kind of light or luminescence that they could find. Then they would gather together all of these different, singular observations and try to reason out some general principles concerning light. Bacon and his contemporaries called these singular observations "particulars," and they formed the core of the Baconian method as it evolved in the seventeenth century. By accumulating and examining particulars, Baconian philosophers sought to establish universals.

If someone examined all of those particular instances of illumination or luminescence, they would start to form ideas about the universal or general nature of light. For example, sunlight, moonlight, and candlelight are all different colors and yet all are examples of light; therefore, one can conclude that light exists in different colors and intensities. They would discover that sunlight and firelight are warm but moonlight is not, and if they examined fireflies they would notice that their luminescence also does not produce heat, so they might conclude that light can exist without heat, suggesting that these are two different phenomena. Their observations might tell them that light travels very quickly across long distances, so perhaps they conclude that light must travel in the most efficient way possible, which would be in straight lines, but by making observations in both air and water they might realize that light moves at different speeds in different mediums. They could keep doing this, drawing on ever more observations to refine their conclusions until they arrive at what they suspect is a series of universal or general principles that describe the nature of light.

Because Bacon's inductive method depends on the observation of particulars, it demands that its practitioners embrace empiricism as a central methodology. The Baconian philosopher believes that they can understand nature only if they observe it over and over again, in different contexts and situations. At the same time, because the key to this method lies in the accumulation of particulars (in other words, it works best when we have more observations to draw upon rather than fewer) it lends itself to collaboration, where multiple people make observations and then share them with each other.

There is another aspect of this focus on particulars that relates specifically to experimentation. Because the inductive method operates from singular experiences or observations, it is well suited to a methodology in which an experimentalist makes different trials or experiments in which they alter certain variables or circumstances and then observe the results. Each singular experiment is useful because it is in the accumulation of experimental results that general principles might be found. While Aristotle's philosophy rejected experiment because it interfered with the ordinary workings of nature, the Baconian method makes singular experiments potentially very useful.

This focus on particulars, however, can also create problems, something that Bacon himself acknowledged. Some events that happen in nature are one-offs, exceptional or rare occurrences that do not reflect universal or general principles. For example, very rarely an animal is born with two heads, the sort of thing that Aristotle would have called an accident to differentiate it from the ordinary course of nature. If someone uses inductive reasoning to determine how nature usually works, however, what do they do with particulars that deviate from an apparent norm? Do they incorporate the occasional two-headed animal in their observations and allow those particulars to change their understanding of general or universal principles? Or do they ignore these "accidents" and remove them from consideration? These are not easy questions to answer, especially if someone is developing a new methodology from the ground up.

Bacon's answer to this problem was fairly simple. Accumulate everything, he argued, and the outliers will eventually drop out of consideration. We cannot know that something we observe is a singular deviation until we accumulate many different instances of the same phenomenon and compare them to one another. But once we do that, we can see quite quickly when one or two singular events deviate from what appears to be the norm. In fact, those abnormalities might actually help us because, in deviating from the way nature normally works, they highlight and show us exactly what "normal" means. This makes sense and works for most natural phenomena that we might want to study. Again, however, there are limitations. For example, there are phenomena in nature that are only rarely observed or encountered. Different examples of light are literally everywhere, but what about an exploding volcano or a comet that becomes visible in the skies only for a few days? There may not be any other examples of these phenomena to compare with your observations, and so you may never know if what you have observed is typical or not.

In spite of these potential problems, Bacon's ideas about empiricism and the accumulation of particulars became increasingly popular through the seventeenth century. He introduced his method in his *Novum Organum* (1620), which he presented as an explicit refutation of Aristotelian natural philosophy. In the later *New Atlantis* (1627), Bacon described an imaginary utopia governed solely by his philosophical, religious, and moral ideas. His philosophy had considerable influence on institutions like the Royal Society of London and in time the Baconian method was praised by prominent thinkers in the mid-eighteenth century, during the Enlightenment. They saw Bacon's ideas as an example of the kind of rational and skeptical inquiry that they themselves wanted to encourage. Whether we want to view Bacon as "the father of the scientific method" or not, he clearly played an important role in the rise of the "new science."

It's Not Easy Being an Experimentalist

Right now, as you are reading this, scientists somewhere in the world are conducting experiments in a laboratory or out in the field. Maybe they achieve a result that they want to share with other scientists, so they publish their findings. But how do the people reading about these experiments know that they actually happened? Why should they trust those claims or results when they weren't present to see the experiment for themselves? Maybe those scientists are lying or making up stories. *Maybe they never performed that experiment at all!*

That last statement might be a little extreme, and yet, why shouldn't we question not just the results achieved by experimental scientists but the very existence of the experiments themselves? How do you or I know that those experiments happened at all? How does anyone? This is not a question posed very often in science today because the modern, collaborative, and global community of scientists employs various methods to reassure one another that they really have done the work that they claim to have done. But this was less easy to accomplish some 400 years ago when experimentation was not yet a consistent part of natural philosophy. In fact, early modern philosophers and experimentalists struggled with basic yet profound questions about trust and credibility when it came to the experimental study of nature, and the methodologies and strategies that they came up with in the seventeenth century helped to shape the modern methods of experimental science that we use today.

How do scientists today reassure one another that their claims and results can be trusted? There are actually many layers of credibility in place. For example, there is credentialing. We expect that the people doing science have the appropriate credentials, usually signified by an advanced degree awarded by a credible institution of higher education. If someone only has an undergraduate degree in, say, history but claims to have proved mathematically that black holes are gateways to parallel dimensions, we would probably raise a skeptical eyebrow and wait for the people with doctorates in astrophysics to weigh in. Another layer of credibility involves replication. It is much easier to persuade someone to believe you if they can replicate your experiment and get the same results. This is one reason why scientists today describe their methods in detail when they publish their work: So other scientists can follow what they have done and hopefully reproduce the same findings in their own labs. There are also processes in place like peer review, in which knowledgeable experts evaluate claims and methodologies before they appear in print, as well as the fact that modern science is a largely collaborative enterprise in which groups of researchers support (but also scrutinize) each other. It seems, then, that modern science has incorporated a range of strategies designed to reassure scientists about the work that others are doing. The same was not always true in the seventeenth century. Back then, experimentalists faced a truly significant challenge: How to persuade others to believe their claims.

Historians of science such as Steven Shapin and Simon Schaffer have argued that one important factor in assigning or creating credibility was social status. Early modern European society, like every other society in human history, had complex standards and rules that governed how different people interacted with one another, rules that often revolved around the question of status or rank. In this period in Europe, the figure of the gentleman was particularly important. The gentleman was a man of means and status: well-educated, worldly, and above all, honorable. A gentleman was always trustworthy. He never lied or exaggerated (his sense of honor made that impossible) which meant that his claims were treated as the simple truth. To suggest otherwise – to suggest that a gentleman was not being truthful – was the same thing as suggesting that he was not honorable, and that was grounds for serious repercussions. Duels were fought and men were killed over questions of honor in early modern Europe.

If we apply this way of thinking to the practice of experimentalism, a gentleman could perform an experiment, report its results, and, generally, be believed. Modern science has substituted other markers of credibility, such as academic credentials, for social status or rank, but the effect is often the same. We trust someone because we have decided that they are trustworthy, a decision made because of *who they are*. This is why the English philosopher Robert Boyle (1627–91) could conduct experiments by himself and then report back to his peers that he had seen a particular result – as a gentleman, his word was honorable and therefore trustworthy. Boyle's peers might disagree with his interpretation of that result, but the result itself was rarely in question, which is important (Figure 6.2).

What if you were not a gentleman, however? This was a problem encountered by a number of natural philosophers in this period, including people like Galileo and Descartes. Descartes did not perform complex experiments – much of his work was based in logic and theory – and so he did not necessarily need to rely on social status to persuade people to trust him. In a sense, his logic spoke for itself and either persuaded people or did not. Galileo faced a more difficult

Figure 6.2 George Vertue's 1739 engraving of Robert Boyle.
Photo by SSPL/Getty Images

problem. He truly did have to convince and persuade large numbers of people that he had seen what he claimed to have seen. This is why Galileo needed a patron, which he found in the Grand Duke of Tuscany. In a sense, the honor and trustworthiness of Cosimo II de' Medici was transferred to Galileo, his client, through the formal arrangement of patronage. As a client of the Medici family, Galileo's claims became inherently more trustworthy and persuasive than they were on their own.

If social status and rank could confer legitimacy and credibility to those performing experiments, their effect could be multiplied by adding more individuals of status. While claims made by Robert Boyle about the results of his work with his air-pump, for example, were deemed trustworthy because Boyle was a gentleman, they could gain even more credibility if other people witnessed that work for themselves. In fact, witnessing became another important methodology that early modern experimentalists came to depend upon. If you could conduct an experiment in front of witnesses who were themselves trustworthy and credible people, then your claims became even stronger as a result. Those witnesses, if they were gentlemen or otherwise trustworthy, could corroborate your account and reassure others that something had actually occurred.

This was especially important in this period because replicating experiments was extremely difficult. The modern scientist does not necessarily need eyewitnesses to her experiments; instead, she can describe her methods and depend on other people in other labs to replicate her work and thereby support her claims. This is only possible today, however, because so much scientific equipment is standardized; the glassware that a scientist uses in India is probably identical to the glassware used in Norway or Kenya or Brazil. While not everyone has equal access to sophisticated equipment or technology, the basic apparatus of modern science is pretty much universal, which means that scientists can more easily and confidently replicate each other's work. But in the seventeenth century, even basic equipment could differ widely from individual to individual. The glass used by someone in London might be very different in thickness or composition from that used by someone in Paris or Madrid or Venice, making replication extremely difficult. Early modern experimentalists themselves acknowledged that this was a problem, but it would be decades, if not longer, before universal standards were adopted in experimental natural philosophy.

With replication difficult to achieve, eyewitnesses become important. Even if someone cannot reproduce Boyle's work with the air pump, they can be reassured that his results are accurate because multiple people observed and reported them. We might consider this an early form of peer review, except that credible witnesses were not necessarily, or even usually, knowledgeable experts. Many were useful only for their inherent credibility, not their expertise. If Lord So-and-So was asked to observe one of Boyle's many experiments involving his air-pump, he *might* have enough understanding about the physical properties of the air to comment knowledgeably about Boyle's results, but it was much more likely that his expertise was inferior to that of Boyle himself. His contribution to the philosophical endeavor was primarily social, in that his presence during the experiment could act as a tacit or silent means of certifying Boyle's claims, rather than intellectual.

When Boyle required the intellectual or expert opinion of his peers, he could take his air-pump to the meeting rooms of the Royal Society of London and conduct his trials there, in front of other philosophers. Or he could publish his results and thereby present them to scholars across Europe who were part of the "republic of letters," the informal but important network of educated people who corresponded back and forth with one another all across Europe and beyond. Some of these people might then critique or support Boyle's claims in correspondence and publications of their own. Again, it was a system reminiscent of the state of science today, though feedback from one's peers is likely to arrive much more quickly now than in Boyle's time.

We have only just started to examine what were actually some very deep problems with the practice of experiment in early modern Europe. Much more can be said about this, and other historians have already done so. For now, it is important to understand that everything from Bacon's focus on singular observations to Boyle's solitary experiments demonstrate two things: First, that the practice of natural philosophy experienced profound changes in the seventeenth century; and second, that those changes created both new opportunities and new problems for early modern naturalists. Those opportunities and challenges were not confined to mainstream or highly visible kinds of natural philosophy, however. They also affected practices like alchemy, an art with a long and controversial history.

The Theory and Practice of Alchemy

Despite its mysterious and sometimes outlandish reputation, at its core alchemy was concerned with understanding and manipulating matter. Its practitioners wanted to puzzle out the secrets of the physical world and use them to change or alter different substances, including the human body. Sometimes called "the Noble Art," alchemy has a history that stretches back beyond Greek antiquity, and few endeavors have inspired more mystery, speculation, and excitement than the extraordinary claims made by alchemists over the centuries. A quick search on Google reveals that self-identified alchemists still exist today, many of them pursuing the same secrets and goals as their premodern forebears.

Alchemy may seem like a strange subject for a chapter on experimentalism, though fundamentally it was a hands-on endeavor; someone did not practice alchemy by sitting back and reading books, but by doing and making. In this way, alchemy lends itself to the practices and methodologies of experimentalism as they emerged in this period. There are other reasons why alchemy makes sense in this context, too. Beginning in the eighteenth century, natural philosophers and other scholars argued that the rationalism and rigor that were thought to be integral to the experimental method were at odds with the vague mysticism and unsubstantiated claims of the alchemist. In other words, the prevailing opinion was that proper, rational experimentalism was incompatible with alchemy. Beginning in the twentieth century, however, historians began to study – as much as possible – what premodern alchemists were actually doing, by sifting patiently through alchemical works and even replicating alchemical processes. Other historians discovered that some of the same individuals who advocated for experimental methodologies were alchemists as well, and that they clearly saw experimentation as not just possible in alchemical work but essential to it.

The remainder of this chapter connects notions about experience and experiment with the theories and practices of premodern alchemy. Doing so helps us understand what experimental life looked like in the "real world" of early modern laboratories and workspaces, but also introduces an ancient discipline that sought to reveal and manipulate the very foundations of nature itself. Alchemy offers a fascinating window on how the "new science" of the seventeenth century influenced and changed prevailing ideas. Transmutational alchemy – that is, alchemical work focused on changing one substance into another, as when lead becomes gold – suffered serious blows to its credibility over the course of the seventeenth century, finally leading to its deliberate separation from the "science" of chemistry in the eighteenth century. In turn, alchemy's demise as a respectable discipline of study will lead into the next and final chapter, which examines how the relationships between religion, magic, and science changed at the dawn of the Enlightenment.

If we want to talk about alchemical practice in the context of premodern ideas, it might be a better idea to talk first about *chymistry*, an early modern word that has been taken up by modern historians such as Lawrence Principe and William Newman as a way to connect alchemical practices with what we today call "chemistry." In fact, the two sets of practices were not easily separated in the past: When early modern people talked about "chymistry" they were referring to a set of methodologies and ideas that concerned the study and manipulation of

Figure 6.3 "The Alchemist" by Philipp Galle, 1558.
Photo by Fine Art Images/Heritage Images/Getty Images

matter, which might involve transmuting metals (something we think of as alchemy) or analyzing the composition of a substance (which to us looks more like chemistry). If a distinction needed to be made between different kinds of processes or activities, premodern people would sometimes refer to *chrysopoeia* as the process of transmutation; the word comes from the Greek and refers to the making or seeking of gold. A chymist could pursue chrysopoeia if he or she wanted to transmute metals, or they could be involved in the production of medicines or dyes or other products of chymistry. Many chymists engaged in all of these activities in one form or another (Figure 6.3).

A great deal of early modern alchemical work revolved around the manipulation or transformation of metals, leading many to wonder where and how metals originated. Some, including Aristotle, believed that metals grew underground in a process not unlike the way that plants grew on the surface. Aristotle in particular had taught that all metals were composed of two principles: Mercury and Sulphur. Not unlike the *tria prima* proposed by Paracelsus, these principles were not literal or physical substances; instead, Aristotle envisioned them as similar to the four classical elements that combined to make up the world. Different metals had these two principles in different proportions, which dictated the properties of each metal. The most perfect metals, like gold and silver, possessed the most perfect proportion of the two principles. The idea that all metals were composed of the same two principles was central to alchemical theory because it implied that the alchemist could change one metal into another by altering the relative proportions of Mercury and Sulphur – lead or tin could become gold if the alchemist knew how to manipulate those two principles.

The Aristotelian conception of metals remained popular well into the Middle Ages and continued to play an important role in shaping ideas during the heyday of premodern alchemy in the sixteenth and seventeenth centuries. How exactly one might rearrange the relative proportions of Sulphur and Mercury in a given piece of metal was a serious question, however. Most alchemists believed that transmutation required the presence of the *Philosopher's Stone*. In basic alchemical theory, the Stone is a substance that can transmute metals, usually by physical contact, though some alchemists also believed that the Stone could cure disease, extend human life, and even permit communication with angels. Few substances in human history have encouraged the awe and desperation that so often accompanied the pursuit of the Philosopher's Stone. Kings and emperors demanded it, and countless individuals worked for entire lifetimes to produce it. More than a few lost their lives in the pursuit, sometimes poisoning themselves unwittingly with heavy metals and other toxic substances, sometimes as punishment for failing to deliver what they had promised.

Interestingly, the history of alchemy includes female practitioners as well as male. The historian Tara Nummedal has studied the case of Anna Zieglerin (c. 1550–75) who, alongside her husband Heinrich Schombach and their friend Philipp Sömmering, performed alchemical work for Duke Julius of Braunschweig-Wolfenbüttel in present-day Germany. Anna had some truly unusual ideas – for example, she described a red oil called "Lion's Blood" that could make trees bear fruit in winter, cure "great, vicious illnesses," and allow women to beget as many children as they desired. She also claimed to have a secret relationship with a mysterious individual known as Count Carl, the illegitimate son of Paracelsus, and that with him she would produce a new race of perfect humans. Her husband Heinrich seemed resigned to this surprising fact; by all accounts his marriage to Anna was not a happy one.

In spite of these outlandish tales, it is clear from the written works that survive that Anna was a practicing alchemist. Nor was she the only woman to practice alchemy in this period; Nummedal has also studied cases of other female practitioners, including noblewomen, who were active in the sixteenth and seventeenth centuries. Very likely there were more, but few accounts exist today – many practitioners, both male and female, did not leave written evidence of their work. The only reason we know Anna Zieglerin's story is that, alongside Heinrich and Philipp, she was placed on trial for various crimes against Duke Julius and his family, before all three were tortured and executed in 1575.

Though unusual in many respects, Anna's story highlights both the extravagant promises that many alchemists made to their patrons and the potential consequences of failure. Many early modern recipes for the Philosopher's Stone required the alchemist to begin with significant quantities of expensive materials including gold, silver, or mercury. Anna's recipe for "Lion's Blood" called for rubies as an ingredient. A wealthy patron was expected to provide these materials, sometimes in massive quantities, but always with the expectation that they would reap even larger returns once the Stone was perfected. More than one unscrupulous alchemist simply disappeared once they had been handed these precious materials, and Anna and her confederates were far from the only practitioners to face severe, or even deadly punishment for failing to deliver what they had promised.

What, exactly, was this substance that inspired such grandiose claims and brutal consequences? Though it was usually called "the Stone," it was not necessarily a stone at all. Some reported that it was a liquid or paste, while others produced grains like sand or pieces of crystal or rock. It was usually white or red in color, and often said to be heavier than its appearance might suggest. Transmutation took place when a metal was heated until it melted and then a small piece of the Stone was added to, or "projected" onto, the molten metal. The change was usually instantaneous.

You might note that I am talking about the Stone as if it was actually produced by alchemists, and transmutation as if it really occurred. This is how premodern alchemists talked to one another, and it highlights a peculiar paradox. Earlier in this chapter we examined what

Figure 6.4 The alchemist in his laboratory. From Michael Maier, *Tripus Aureus*, 1618.
Photo by Hulton Archive/Getty Images

"experience" looked like to premodern people, as well as problems like replication and credibility which experimentalists had to address or overcome in making the results of their work public. The paradox here is that some of the same people who championed these new and evolving methodologies were also committed alchemists, and in their alchemical work they were sometimes forced to adopt different standards of replication and credibility than those they applied to other experimental work (Figure 6.4).

Generally speaking, alchemy was what we might consider a hidden or obscured art. Its methods and processes were not intended for general consumption. In fact, alchemical writers took great pains to ensure that only the most skilled and knowledgeable practitioners could hope to understand and replicate alchemical processes. This secrecy was necessary because the power of the Philosopher's Stone was simply too great to be allowed to rest in the hands of just anyone. Only the deserving could be allowed to know the secrets of alchemy, so virtually all alchemical authors obscured descriptions of alchemical work and the Stone itself with elaborate metaphors, poetry, and imagery.

Consider the following example, taken from the beautiful and confusing *Atalanta fugiens* of Michael Maier (1568–1622). The book is made up of fifty "discourses," each of which presents an image or emblem accompanied by poetry, prose, and musical notes. Each discourse illustrates a single step in the production of the Philosopher's Stone, but they are not necessarily arranged in the correct order. The task for the aspiring alchemist is to decode the meanings hidden in each discourse and then rearrange them in the correct sequence that, when followed, will produce the Philosopher's Stone. Discourse number twenty-four begins with an image titled, *Regem lupus voravit, & vitæ crematus reddidit*, which translates to, "A wolf devoured the king, and when burned restored him to life." In the accompanying image we see a male figure wearing a crown, lying on the ground while a wolf crouches over him, devouring his body. In the background we see a large bonfire, at the center of which burns that same wolf, and walking from the flames is the king, restored to life (Figure 6.5).

This is actually one of the easier images to decipher in the *Atalanta fugiens*, though doing so requires that we know a few basic facts about both alchemical symbolism and chymical processes. The Grey Wolf usually indicated the element antimony, a grayish or silvery substance that, like a wolf, was voracious in its ability to "devour" or combine with almost every known metal. The king in this case is gold, the highest metal. The image, then, instructs the alchemist to combine gold (already produced in an earlier step of the process) with antimony and then heat the resulting mixture. They will be left with a purified and "revived" form of gold that can then be used in the next steps. Note that there are no indications here as to the amounts of gold or antimony required, nor any instructions as to how long to heat the mixture or at what temperature. These are details that the alchemist would have to uncover on their own, or that perhaps they already knew from previous attempts. Frustratingly, however, if they did not perform this particular step correctly, they might not know it until weeks or months later when the whole process failed to produce the Stone.

It is challenging to reconcile these methodologies and practices with the emerging fashion for experimentalism that took place in the seventeenth century. Alchemy was a collaborative enterprise only in the

Figure 6.5 An engraving from Michael Maier's *Atalanta fugiens* (1618): "A wolf devoured the king, and when burned restored him to life."
Photo by Photo 12/ Universal Images Group via Getty Images

loosest sense. While many alchemists corresponded with one another, passing along hints or asking questions about the pursuit of the Philosopher's Stone, for the most part this was a solitary and secretive endeavor. If someone discovered how to produce the Stone, it is very unlikely they would want to share that with anyone else. Replication was also a particular problem for alchemists, more so even than for their contemporaries working in other branches of natural philosophy. The vague and cryptic instructions that usually characterized alchemical recipes virtually guaranteed that attempts to replicate a particular process would be extremely difficult. Add this to the lack of standardized equipment and methods, and even if someone did manage to produce the Stone, it was highly unlikely that their glassware and other equipment was identical to that used by others. Even small differences in thickness or quality could affect results dramatically.

Credibility was another persistent problem for early modern alchemists. In spite of the many obstacles that complicated the pursuit of the Philosopher's Stone, surprising numbers of alchemists claimed to have produced it. We cannot know how many of these claims were false, but given how difficult the process was, it is likely that many were. This could (and did) lead to accusations that all alchemists were a collection of lying frauds, but the historian of alchemy Lawrence Principe has argued that a better way to understand these claims is not as examples of fraud, but of what he calls "chymical optimism." According to alchemical (or chymical) theory the Philosopher's Stone was a real object that *could be* created, but it was extraordinarily difficult to do so. That an alchemist failed in their attempts to create the Stone was not proof that it did not exist; on the contrary, the dedicated alchemist was much more likely to assume that they had made an error at some point in the process. In this context, claims to have created the Stone can be understood not as reliable indicators of something that actually happened but as optimistic statements about what could, should, or would eventually occur.

Nor was credibility an issue only among the skeptical; even alchemists themselves were wary of claims that seemed too good to be true. A common means of acquiring recipes for the Stone or other alchemical preparations was through correspondence with other practitioners, and some, like Robert Boyle, devoted considerable time and, in some cases, considerable wealth to fostering these connections. It was not uncommon for the promised recipe to prove inadequate or nonexistent, but there were darker, more concerning possibilities as well. Boyle, for example, recorded an instance in which he himself witnessed the transmutation of lead into gold, conducted by a visitor to whom he had been introduced but did not know. Toward the end of his life he confessed to his friend Gilbert Burnet (1643–1715), bishop of Salisbury, that this and other incidents troubled him greatly. He believed that he had witnessed a true transmutation and that the Stone was real, but he was uncertain whether he could trust this visitor. Perhaps he, or the other mysterious alchemical adepts who had offered to share with Boyle their substances and discoveries, had achieved these great works through the aid of demons. By accepting their aid, was Boyle imperiling his own soul? He was far from alone in wondering whether a desperate alchemist might turn to demonic or diabolical aid after failing to produce the Stone – the Jesuit Athanasius

Kircher warned of just such a possibility, as did others going back to the Middle Ages.

Ultimately, however, Boyle's misgivings did not prevent him from pursuing transmutational alchemy. This despite the fact that, in many respects, alchemy did not lend itself well to the experimental methodologies that emerged in the seventeenth century, of which Boyle himself was a fierce proponent. This might be explained in part by the fact that alchemy and its methods were many centuries older than the experimental philosophies of Francis Bacon or the Royal Society, making the one difficult to reconcile with the other. The difficulty in reconciling these two systems was one reason why leading chymists eventually created a new disciplinary understanding of chymistry that embraced the experimental rigor of the "new science" while distancing themselves from the increasingly problematic realm of transmutational alchemy.

Alchemy was not defeated by the rise of a new and rational science, however. Many careful and skeptical proponents of the "new science" continued to pursue metallic transmutation into the early decades of the eighteenth century. In fact, the person hailed by historians as the preeminent early modern scientist was also a dedicated alchemist. That person is Isaac Newton (1643–1727), and we close this chapter by examining the role that alchemy played in shaping his natural philosophy.

Newton the Alchemist

Historians of science tend to view Isaac Newton as the ultimate product of the "Scientific Revolution" that began with Nicolaus Copernicus (Figure 6.6). His philosophical work integrated many of the innovations described in this book and tied them together into a single coherent system that persisted until the early twentieth century when it was challenged by the theories of Albert Einstein (1879–1955). But while Newton has long been hailed as one of the greatest scientists in history, he was also a practicing alchemist and a deeply pious individual with a strong interest in the study of religion. Unfortunately, those aspects of his thought remained largely unknown and ignored after his death, only resurfacing in the mid-twentieth century when some of his alchemical papers came to light. The British economist John Maynard Keynes (1883–1946), after purchasing some of Newton's alchemical

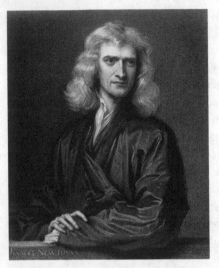

Figure 6.6 Portrait of Isaac Newton, 1689.
Photo by SSPL/Getty Images

notes, famously wrote in 1942 that "Newton was not the first of the age of reason. He was the last of the magicians." It was only after this realization that Newton's scientific works were reevaluated in the context of his other interests, though with considerable reluctance on the part of many historians of science.

Newton was born in 1643 and educated in the last vestiges of Aristotelian philosophy that still persisted in universities of the time. He was also introduced to the mechanical philosophies proposed decades earlier by Gassendi and Descartes, the heliocentric cosmology popularized by Galileo and Kepler, and the Baconian method of inductive philosophy. In addition, the available evidence suggests that from a relatively young age Newton was an avid alchemist with an interest in the properties of matter, though we know much less about his alchemical studies than we do about his more "mainstream" work. In part this is because a great deal of alchemical material was destroyed during a fire in Newton's laboratory (caused, according to one story, by Newton's dog), and in part because Newton appears to have kept much of his alchemical work a secret. Given that the reputation of transmutational alchemy was in decline toward the end of the seventeenth century, this desire for secrecy is not surprising.

Beginning in the 1970s, the historian of science Betty Jo Teeter Dobbs was one of the first to look seriously at Newton's alchemical studies and suggest that they played a role in shaping his natural philosophical and mathematical achievements. Her claims were met with skepticism and even hostility from historians who refused to believe that something as "unscientific" as alchemy could have played a formative role in some of the most famous scientific work of the premodern era. Dobbs demonstrated, however, that one of Newton's earliest concerns was with the problem of inert or inactive matter, something that preoccupied many philosophers in the seventeenth century. His notes and other writings indicate that he was troubled by the idea that matter was essentially dead or inert. If this were true, then explanations for complexity in nature would have to rely on either purely mechanical causes, such as the movement and collisions of small pieces of matter, or direct intervention from some outside cause. The latter explanation was not appealing for several reasons, not least because it suggested that God Himself might be required to move all matter all the time. Purely mechanical causes, on the other hand, seemed insufficient to explain especially tricky phenomena like gravity and magnetism. Newton, then, began by considering the possibility that matter was imbued with some kind of active principle or attractive virtue that could explain why particular phenomena occur in nature.

By carefully examining Newton's surviving manuscripts, Dobbs established that this interest in active principles came directly from his alchemical studies. Many alchemists taught that matter was not inert but possessed some vivifying or "vegetative" principle that allowed it to behave in ways similar to living or organic substances. In fact, vegetative and organic metaphors were common in alchemical texts that described substances as being born or generated in the chemical flask or growing like plants or trees when exposed to ideal conditions. When Newton first considered the idea that matter might be imbued with these active principles, he was transferring an old idea from alchemical literature to the new paradigm of the mechanical philosophies.

After more than twenty years of studying Newton's work, Dobbs concluded that his primary concern was establishing not just that God had created the universe, but also that He acted continuously within it. Though Newton's ideas evolved and changed over his lifetime, his belief that God was an active part of the world never wavered. At various points, Newton considered the possibility that light or gravity were examples of God's activity in the universe, and also speculated that the active and animating principle inherent to matter might be proof of the same thing. Thus, we find Newton struggling with the same philosophical problems confronted by most advocates of the mechanical philosophies – the presence and role of God, the idea of dead or inert matter – and ultimately reaching into his alchemical studies for what he saw as compelling answers.

This has numerous connections to the aspects of Newton's work that most of us know best. If one wanted to simplify Newton's great achievement in physics and cosmology, they might say that he took the mechanical, materialist philosophies of the seventeenth century and added to them the concept of an immaterial and invisible force. He first described this idea in the work for which Newton is best known today: the *Philosophiae naturalis principia mathematica* or "The Mathematical Principles of Natural Philosophy," usually known simply as the *Principia*. Newton proved mathematically that gravity was a universal force that acted on all matter according to the same law, meaning that the same force that caused an apple to fall from a tree also held the Moon in its orbit around the Earth. This was revolutionary in itself, as it provided a simple and elegant explanation for many kinds of physical motion and tied together the physics operating in both the terrestrial and celestial realms. The addition of an invisible, occult force also allowed Newton to solve one of the most glaring problems with the mechanical philosophies: namely, proof for the existence and presence of God. For Newton, gravity was the unseen hand of God at work in the universe, holding everything together and keeping it in motion.

Newton's fixation on divine activity in nature was no accident. More than either his alchemical work or his innovations in physics and mathematics, religion and theology actually occupy the majority of Newton's surviving papers. He devoted considerable study to the Biblical account of the Temple of Solomon, considered by many in early modern Europe to have contained numerous occult secrets encoded in its architecture and proportions. He was also fascinated by Biblical prophecies, particularly those related to the Book of Revelation and the end of the world. Interestingly, however, it seems that Newton was not an orthodox Anglican at all; instead, he privately followed the doctrines of *Arianism*, which denied the reality of the divine Trinity accepted by all orthodox branches of Christianity. Descended from one of the earliest Christian sects, Arianism taught that Christ is not the same person as God but is instead a semidivine intermediary between God and humanity. Had these views been known at the time, Newton would have faced harsh condemnation from his peers and his public reputation would have suffered. He succeeded in keeping his unorthodox Christianity a secret, however, and it was discovered only after his death.

In Isaac Newton we find a personification of the rich and colorful tapestry of seventeenth-century ideas. He tied together the mechanical philosophies and a heliocentric cosmology in a single elegant system that borrowed from alchemical theory to establish a role for God in the universe. At first glance, the world described in Newton's work looks entirely different from that occupied by people who had lived a hundred years earlier, and yet some important things remained the same. There was no question for Newton that God acted in the universe and that the study of nature would reveal that activity. Likewise, he believed that the manipulation of nature's occult and hidden parts was a critical part of natural philosophy. We should not see Keynes's characterization of Newton as "the last of the magicians" as an indictment, but as a celebration of the brilliance with which he took hold of the whole panoply of early modern ideas and channeled them into a beautiful and compelling vision of the universe.

In hindsight, however, we can also understand Newton as a transitional figure. The educated classes of the eighteenth century cast him as a paragon of rationality and an exemplar of modern science. Stripped of its religious and occult connotations, Newton's physics became a model for Enlightenment ideas, thanks to its emphasis on mathematics and the universal nature of its claims. Yet, the Newton celebrated by many in the Enlightenment was only part of the complex person we are now beginning to understand.

The careful editing of Newtonian ideas and their place in the eighteenth century takes us into our next and final chapter, which explores how the many strands we have followed throughout this book came together at a point in time that historians have long considered the beginning of Western modernity. Newton's singular vision of the cosmos, which incorporated mathematics, occult forces, active principles, and theology, may have been a complex and multilayered expression of early modern ideas, but in the Enlightenment skeptics and reformers alike teased apart the realms of magic, religion, and science in their attempts to fashion a new understanding of their world.

7 A New World?
The Dawn of the Enlightenment

Lawrence Principe has called the sixteenth and seventeenth centuries the "golden age" of alchemy. Scholarly interest in alchemical pursuits was at an all-time high, and wealthy patrons across Europe were eager to secure alchemists as clients. This was also a good time for chymistry more broadly, which saw its status as a respectable discipline improve to the point where it became part of the curriculum at many universities, often as an ancillary branch of medicine. This was a boon to the increasing numbers of chymical physicians (many inspired by the ideas of Paracelsus) and to those, like Robert Boyle, who believed that chymistry offered a powerful and useful way of investigating nature. As chymistry gained more academic respectability, however, there was a curious reversal in the fortunes of transmutational alchemy. Supporters and practitioners of chymistry began to distance themselves from the idea of metallic transmutation. Ultimately, the many failures of transmutational alchemists to produce the Stone, combined with increasing accusations of fraud, eroded confidence in *chrysopoeia*, and by the early decades of the eighteenth century many alchemists were viewed as hucksters and quacks, making grandiose claims while taking money from the gullible and the desperate.

In 1718, the Dutch professor of medicine and chemistry Herman Boerhaave (1668–1738) used his new position at Leiden University to call for a reform of chymistry. He denounced the false claims made by transmutational alchemists and urged his colleagues to purge these "errors" from the practice and teaching of chymistry. Thus began the swift and final demise of transmutational alchemy as a respectable field of study. Alchemists became figures of ridicule in plays and stories while, in the universities, academic scholars systematically excised *chrysopoeia* from chymical curricula. As chymistry became an academic subject in its own right, no longer a mere adjunct to medicine, its supporters worked to boost its respectability by distancing it from the problematic associations carried by transmutational alchemy.

189

The ignoble demise of alchemy is an important part of the narrative in this book. For one thing, it demonstrates how quickly sentiments changed. After a colorful history that had lasted for more than 2,000 years, it took perhaps one or two generations for transmutational alchemy to fall from grace. At the same time, this episode also demonstrates how philosophers and naturalists engineered a clear division between "respectable" knowledge on the one hand and superstition and error on the other. Nor was it just chymistry that they purged of these unwelcome attributes. The study of nature underwent a profound transformation in the first half of the eighteenth century, as part of which naturalists deliberately separated occult traditions from the "rational" and respectable pursuit of natural philosophy. This was the moment when magic was cast out of science and became the realm of superstition and trickery we still associate with it today.

This process was not a simple one, and nor was it wholly complete. Supporters of the "new science" quietly adopted elements of occult philosophies to strengthen their systems of inquiry, meaning that modern science owes at least part of its existence to the magical ideas explored throughout this book. The rise of Newtonian science around the beginning of the eighteenth century also played an important role in this adoption of occult principles. All of this played out against the backdrop of the Enlightenment, which introduced profound changes to European society. By the end of our narrative around the year 1750, we will find a Europe that looked, in at least some respects, recognizably modern.

The Age of Reason

Broadly speaking, the Enlightenment was a cultural and intellectual movement that occupied much of the eighteenth century. It is also sometimes called "the Age of Reason" because reason and rationality are seen as central to the goals and consequences of the Enlightenment. There was a general movement away from traditional systems of learning and governance, something that had already started in the previous century but which accelerated considerably in the eighteenth. Writing toward the end of the century and reflecting on the changes that had taken place in the previous decades, the German philosopher Immanuel Kant (1724–1804) borrowed the phrase *Sapere aude*, or "Dare to know," from the classical poet Horace and claimed this as the motto of the Enlightenment. For Kant, this expressed his own belief that human reason should govern progress and change in society, and in this respect Kant was very much a follower of the central ideals of the Enlightenment.

It is challenging to talk about the Enlightenment as if it was a singular phenomenon because it looked very different in different places. Some of the most important articulations of Enlightenment ideals originated in France, but other countries experienced the Enlightenment in different ways. Whereas many French thinkers attacked the dogmatic traditions of the Catholic Church and its influence on French society, people living in the German states were generally more interested in reforming the practice and structure of government, and some historians remain uncertain whether the Enlightenment actually took hold in Britain at all. There are enough common elements across different nations, however, to suggest that we can identify some universal beliefs and ideals that defined "the Enlightenment" for most people.

It is widely accepted that this period in European history defined much of what we in the West now understand as "modernity." In other words, the Enlightenment effectively created the idea of the modern West as most of us experience it today. For example, the separation of church and state enshrined in most modern democracies was articulated most forcefully by Enlightenment thinkers, along with ideas about religious tolerance and the importance of individual liberty. Most of these changes were rooted in conscious and deliberate reactions against the *status quo* that had prevailed in Europe for centuries.

As we generally understand it today, to be enlightened is to be modern and open-minded. This is no accident; the individuals at the forefront of the Enlightenment modeled in their own lives a progressive ideal that equated rationality and tolerance with modernity. Of course, this ideal had limits. Notions of tolerance and liberty generally were applied only to white men, and existed in clear opposition to the practice of slavery which still existed in some European colonies during the eighteenth century. Similarly, the famous cry of "Fraternity!" or brotherhood that defined the spirit of the French Revolution at the end of the eighteenth century excluded women and the poor. If the Enlightenment gave us modernity, it also left us with some of the most enduring social problems of the modern era, including racism, sexism, and a persistent lack of respect for the working classes.

Though many people living in the Enlightenment applied its ideals imperfectly, however, those same ideals still represent a profound change in how Europeans understood their own society as well as the wider universe. This was in part a culmination of some of the trends described in previous chapters: for example, the slow but steady rejection of ancient authority and its replacement by innovative methods of inquiry and experimentation. At the same time, the strong connections between natural philosophy, religion, and magic that had persisted for hundreds of years became deeply and irrevocably strained in the eighteenth century. Some of the most outspoken Enlightenment thinkers dismissed both organized religion and magical beliefs as ignorant superstition, even as they quietly integrated elements of earlier magical philosophies and practices into the new and powerful natural philosophy that came to dominate the eighteenth century.

The emphasis on reason in the Enlightenment tended to privilege particular ways of thinking about the world and, in turn, created new institutions and priorities for European society. If someone wanted to argue that the application of reason was crucial to the development of a new and enlightened nation-state, then public education would need to change in order to cultivate a properly rational mindset in that nation's citizens. Disciplines that had already embraced the exercise of reason, such as the physical and mathematical sciences, could now act as important exemplars for other disciplines, meaning that educated people began to emphasize quantitative methods and approaches in fields like biology, chemistry, and anthropology. At the same time, anything that might endanger the exercise of reason, particularly the irrational belief in religious dogma or the divine basis for the monarchy, needed to be minimized or suppressed. To varying degrees, all of these changes happened in different places during the Enlightenment.

Science in the Enlightenment

The creation of the first "scientific" societies such as the Accademia del Cimento in Florence (founded in 1657), the Royal Society of London (founded in 1660), and the Académie Royale des Sciences in France (founded in 1666) created new communities of philosophers and experimentalists in the latter decades of the seventeenth century, and this model of collaborative inquiry and the open exchange of ideas continued to grow into the eighteenth century. These societies did not necessarily operate in the same ways, however. The French monarchs used the expertise of the scholars assembled in their Académie to advance projects of particular interest to the crown and the state, while the Royal Society of London enjoyed the rather disinterested patronage of Charles II at its founding and thereafter received little in the way of formal support from the English crown. Nonetheless, these and similar societies became important hubs for scientific cooperation and innovation in the eighteenth century. Smaller regional societies sprang up across Europe, allowing the formation of new and larger networks of naturalists, philosophers, physicians, and others with an interest in the study of nature.

As the institutions of scientific and philosophical activity expanded, so too did the very concept of "nature," which assumed new and important connotations over the course of the Enlightenment. Dorinda Outram has echoed other historians in pointing out that the educated classes began to treat nature as a kind of "ethical norm" in the eighteenth century. Nature became associated with what was considered "good" or "pure" or "right," and ethical philosophies in the Enlightenment increasingly rejected religious or Scriptural morality in favor of arguments that depended upon natural virtues and qualities that many believed were inherent to all human beings. Jean-Jacques Rousseau (1712–78) famously argued in his work *The Social Contract* that civilization was an artificial construct that tended to bring out the worst qualities in people, and he wrote approvingly of humans who lived without the kinds of civilized society seen in Europe and who, according to Rousseau, possessed an uncorrupted kind of morality that arose from their natural state. While he was sharply critical of what he saw as the many excesses and vices created by civilization, however, Rousseau also believed that the only way to create truly virtuous and moral people was through education founded on reason. Rousseau may have envied those who lived outside of civilization for their natural innocence, but ultimately he saw true morality as belonging only to those educated in the rational ideals of the Enlightenment.

Another important change in how Enlightenment thinkers saw nature was in their increasing emphasis on concepts such as "natural order" and "natural laws." Many in the Enlightenment described the natural world as governed by immutable and rational laws, an idea expressed by René Descartes as early as the 1630s but which was embraced much more widely in the eighteenth century. This was due in part to the work of Isaac Newton, whose *Principia* had demonstrated that someone could describe even the invisible force of gravitation in mathematical terms. Later popularizers of Newton's ideas would claim that Newtonian mechanics proved the existence of an ordered and rational cosmos whose laws were both mathematical and comprehensible.

Some also applied similar ideas to the perceived order of nature, which became a topic of particular interest for biologists and botanists. The Swedish zoologist Carl Linnaeus (1707–78) is best known today for his taxonomies of the plant and animal kingdoms, a decades-long endeavor that Linnaeus saw as expressing in rational terms the order that was innate to Nature itself. He was far from the only person interested in taxonomy, however, which educated people applied in this period to every living thing, including human beings. Human taxonomies were usually shockingly racist by modern standards, as those writing these taxonomies (white men) invariably put themselves at the very top of a racialized hierarchy. These hierarchies, which were seen as natural and thus both "just" or "right" and also devoid of cultural or political bias, became a powerful justification for practices that included the widespread colonization of distant places and the enslavement or disenfranchisement of those deemed inferior to the European male.

Alongside this interest in taxonomies and hierarchies arose some of the first claims that the Earth was significantly older than people had once believed. Traditionally, the age of the Earth had been estimated using sources like the genealogies in Scripture; the bishop James Ussher (1581–1656) had calculated that the first moment of Creation had occurred on the 2nd of October in the year 4004 BCE. These limited chronologies were challenged in the eighteenth century by new forms of evidence, leading some like Georges-Louis Leclerc (1708–88, usually known as the Comte de Buffon) to argue that the Earth was in fact around 75,000 years old. Buffon examined fossils and made geological observations in order to arrive at this figure, which was condemned by theologians in Paris and elsewhere. Nonetheless, these first inklings as to the Earth's extensive history would go on to have a profound impact on later thinkers including Georges Cuvier (1769–1832), Charles Lyell (1797–1875), and Charles Darwin (1809–82).

While these developments preoccupied the educated elite, the average person living in Europe enjoyed greater exposure to scientific innovations and knowledge over the course of the eighteenth century. This period marks the emergence of a new kind of public science, motivated by the belief that the ideal citizen of the Enlightenment should be both well informed and broadly educated. It became increasingly common for scientific societies to stage public demonstrations and educational exhibitions, eventually leading to the widespread integration of scientific and technological spectacles in various forms of entertainment and recreation. Scientific ideas and accomplishments also received greater coverage in newspapers and periodicals, helping to create a more scientifically literate middle class.

In a little over 200 years the Earth had gone from the center of a small cosmos to a solitary speck moving in an enormous universe. Someone born in 1750 probably had grandparents who were alive when Newton published his *Principia* and turned the European understanding of the universe on its head. As quickly as the world was changing, it is no surprise that religion had to run to catch up.

Religion in the Enlightenment

For some 200 years, the general understanding of the Enlightenment has been that it caused profound changes in European religion. Like Kant, the German philosopher Georg Wilhelm Friedrich Hegel (1770–1831) used the power of hindsight to look back on the Enlightenment and describe it for those living in its aftermath. He saw the Enlightenment as a continuation of the Protestant Reformation, but one carried out in the eighteenth century by philosophers rather than theologians. Both movements, for Hegel, had ideals of freedom and individual expression as their cornerstones, just as both had the potential to forge a stronger connection between the individual person and the spiritual realm. But he also believed that Enlightenment thinkers ultimately failed to understand the human value in religion, treating it mainly as a means of social control or as an outgrowth of the study of nature. In spite of that failure, however, Hegel believed that the Enlightenment was, at its core, concerned with religion above all else.

There is no doubt that religion experienced profound changes in the eighteenth century. Some of the best known participants in the Enlightenment, such as Denis Diderot (1713–84) and Voltaire

(1694–1778), wrote scathing attacks on the Catholic Church and called more generally for an end to the political and social influence enjoyed by organized religion. It is no coincidence that both were French; the Enlightenment in France was characterized first and foremost by commentaries on the Catholic Church, many of them sharply critical. Challenges to traditional religious institutions also arose from within, in the form of powerful movements for reform that reflected a desire on the part of many people to leave behind the religious turmoil that had characterized much of the sixteenth and seventeenth centuries. For example, leaders of the movement known as *Pietism*, which arose within German Lutheranism, advocated for a new spiritual focus on the individual's communion with God and the pursuit of a good and simple Christian life. Across the Atlantic Ocean, the First Great Awakening triggered a period of revival that changed Protestant Christianity by introducing a greater emphasis on individual accountability and piety. This was the beginning of evangelical Christianity in North America, and was followed by later Awakenings in the nineteenth and twentieth centuries.

As Pietism and other reform movements appeared in the eighteenth century, they were met by a widespread interest in fostering greater religious tolerance. The impetus, again, was a wish to avoid the brutal wars of religion that had swept across so much of Europe in the preceding centuries. Writers, philosophers, members of the clergy, and heads of state all came together to advocate for a new kind of European society that embraced greater religious tolerance and an end to religious persecution. These ideals also dovetailed neatly with the Enlightenment emphasis on individual liberty and freedom. Though tolerance was far from universal, we find in the eighteenth century the rise of nation-states defined not by religious factionalism but by increasingly secular forms of national identity.

The calls for secularism advanced by some Enlightenment thinkers, as well as the smaller role played by religion in political life, should not lead us to assume that religious faith diminished in the eighteenth century, however. On the contrary, the average person was just as likely to attend church or believe in God at the end of the century as they were at the beginning, though what they heard from the pulpit might have changed significantly in that time. This is worth remembering because the mythology of the Enlightenment created in the nineteenth and twentieth centuries has tended to portray it as the rise of secularism and the end of religious dominance in Europe. It is true that a conscious and deliberate separation appeared between religion and politics in this period, and that more people felt free to declare their public support for materialism or even outright atheism, something that would have been unthinkable a century earlier. Religious belief did not disappear in the Enlightenment, however. It changed, adapted, and evolved in new directions.

The idea that Nature itself was a source of ethical or moral values had profound implications for religion in the eighteenth century. Increasing numbers of philosophers and theologians sought to find a convenient middle ground between these two realms, and the notion of a natural or rational form of religion was seen by some as a necessary alterative to the irrational and superstitious dogma of past religious institutions. In part, this desire led to the rise of *deism*, a religious philosophy which claimed that God existed and had created the universe but no longer interfered in His creation. Deists believed that God's existence could be inferred from observable phenomena like the existence of regular natural laws, but that God was otherwise unknowable. This notion of a distant and largely unknowable God was appealing to many of the "enlightened" because it was easier to reconcile with the increasingly materialist and mechanistic philosophy that prevailed during much of the eighteenth century. The focus of natural philosophy in this period centered on the study of nature's laws, which required nothing more than a passing acknowledgement that those laws were the work of a providential God, if that. Most deists took God's existence as a given and otherwise gave it little thought. Compare this with the efforts of philosophers like Descartes and Gassendi who, a century earlier, had tied themselves into rhetorical knots trying to find a place for God in their mechanical systems. Enlightenment thinkers solved that problem by reconceptualizing God rather than the universe.

Deism did not necessarily overtake other ways of understanding God in the eighteenth century, but it was influential. For one thing, it helped to usher in a revival of *natural theology*, which had existed in one form or another from the Middle Ages but which assumed new importance in the Enlightenment. Natural theology argues that our best (and perhaps only) way of knowing God is by studying nature.

This kind of theology generally rejects other ways of knowing God, particularly direct revelation as found in miraculous events or the revealed word of God recorded in Scripture. Instead, it privileges the exercise of reason (itself a very "enlightened" ideal) and the empirical study of the universe. Natural theology would go on to even greater prominence in the nineteenth and twentieth centuries, both in Europe and the United States, and inspired the rise of creationism and the doctrine of "intelligent design" advocated by some evangelical Christian communities today.

Religious tolerance, sweeping reform movements, and the rise of "reasonable" Christianity and natural theology all combined in the Enlightenment to change both the understanding and the practice of religion. As with science, this reflected a changing conception of the wider world and humanity's place in it. But was there still room in that world for magic?

Magic in the Enlightenment

In the 1970s, the historian Keith Thomas famously described what he called the "disenchantment of nature" as part of a larger argument about popular belief in early modern England. Thomas claimed that popular belief in magic gradually faded away before the year 1700 but that religious belief remained strong and vibrant in communities across England. Most premodern people believed in both; they embraced an eclectic worldview in which both religion and magic were seen as useful and necessary. Thomas, however, saw a distinction between the two. Religion gave people meaning and purpose, while magic was a convenient way to deal with immediate but temporary problems. As early modern people became more self-reliant (due, Thomas suggested, to the influence of a uniquely Protestant ethic) their need for magic diminished. This, combined with the rise of the "new science" and a more rational outlook generally, led to the slow decline of magic and the disenchantment of the world.

When Thomas published these ideas in 1971 they were a revelation for historians. He demonstrated that popular beliefs, not just the writings of the educated elite, offered an important way to understand past ideas. He also suggested that religion and science had not been enemies after all, a radical proposition for the time. His book became required reading in the study of early modern history for the next 30 years, even as historians who came after him slowly chipped away at some of its more sweeping claims.

Scholars in the twentieth century had good reasons to believe the narrative that Thomas presented. For one, it aligned with the narrative constructed by Enlightenment thinkers eager to portray themselves as rational people who had abandoned "vulgar" and ignorant superstition. More recently, however, historians tend to view the idea of "disenchantment" with caution. It seems clear that attitudes toward magic did change in the seventeenth century and that, for much of the eighteenth century, we find numerous people claiming that a belief in magic was irrational, superstitious, and ignorant. Yet, there is evidence that magical beliefs were not swept aside by scientific rigor and a commitment to rationality, as the disenchantment theory would suggest.

A growing amount of historical scholarship now argues that magical beliefs and practices had an important influence on the development of natural philosophy, and that around the beginning of the eighteenth century the educated classes chose to retain some elements of magical systems while rejecting others. Of course, previous chapters in this book have shown repeatedly that magical and occult philosophies had long been central to how people studied the natural world. Consider the hermetic and cabalistic influences on John Dee, Paracelsus's quest for nature's hidden secrets, or Isaac Newton's alchemical experiments. At the same time, the mechanical systems of Gassendi and Descartes, which were dependent on the unseen motion of invisible pieces of matter, presented people in the seventeenth century with occult or hidden explanations for natural phenomena that functioned much like earlier systems that had depended on invisible sympathies or magical forces. Recall the weapon salve, and the way in which explanations for its ability to heal wounds over distances changed over the course of the seventeenth century from immaterial, sympathetic connections and quasi-magnetic forces to invisible particles moving through the air. Nonetheless, it still healed distant wounds. The evolution of explanations for the weapon salve demonstrates that occult activity remained central to natural philosophy. People simply reimagined hermetic sympathies and cabalistic correspondences as invisible corpuscles and atoms.

This shift in how early modern people conceptualized and used occult causes leads us to the work of the historian John Henry, who has suggested that, rather than disenchantment, we should understand the fate of magic as one of *fragmentation* in which philosophers retained some elements of magic and rejected others. The case of transmutational alchemy at the beginning of this chapter illustrates Henry's narrative of fragmentation perfectly. Early modern chymistry included a wide range of different practices and methodologies, including *chrysopoeia*, the pursuit of metallic transmutation. When Boerhaave called for a reformation of chymistry in 1718 he was concerned about the respectability of the discipline, which he saw as endangered by the fraud and trickery of quack alchemists. He knew very well, however, that *chrysopoeia* was only a small part of the larger practice of chymistry, even in the heyday of alchemy in the sixteenth and seventeenth centuries, and he also understood that the discipline of chymistry had already integrated fundamental alchemical ideas and practices into its foundations. The study and transformation of matter, which had been central to alchemical work for hundreds of years, also defined the discipline of chemistry as it emerged in the eighteenth century. Boerhaave's deliberate attempts to draw a line between "respectable" chymical work and the fraudulent practices of transmutational alchemists were therefore not a wholescale rejection of alchemy. Instead, it was a careful repudiation of particular alchemical practices. He, and others like him, tried to establish chymistry as a respectable discipline of academic study by breaking it apart into pieces. They separated and then pruned away its most troublesome elements, leaving behind a set of theories and practices with a deep (but unspoken) debt to alchemy.

Another example of this narrative is the way in which some have divided Isaac Newton's ideas into "respectable" and "contemptible." More than any other natural philosopher, Newton was revered by scholars in the Enlightenment as an exemplar of reason and logic. Many in the Enlightenment were either unaware of Newton's religious and alchemical interests or ignored them. When Sir David Brewster (1781–1868) published the first systematic biography of Newton in 1855, he described his feelings when he first encountered some of Newton's alchemical papers by writing, "We cannot understand how a mind of such power, and so nobly occupied with the abstractions of geometry, and the study of the material world, could stoop to be even the copyist of the most contemptible alchemical poetry, and the

annotator of a work, the obvious production of a fool and a knave."[1]
Brewster's contempt and revulsion are palpable, and reflect the one-
sided view of Newton that rose to prominence during the
Enlightenment. We know now that the Newtonian universe, governed
by rational laws that could be established by logic and mathematics
and seemingly uncomplicated by troublesome religious or theological
issues, was actually the product of Newton's fierce dedication to the
idea of divine intervention and built on a foundation of active or
attractive principles that he imported directly from his alchemical
studies.

Ultimately, magical beliefs and occult systems were already part of
the natural philosophies that proliferated in the Enlightenment. The
educated classes of the eighteenth century adopted Newtonian science
with enthusiasm, unaware or uncaring that its foundations were
rooted firmly in religious and alchemical traditions. At the same time,
they loudly and self-consciously removed obvious traces of religion
and magic from the wider study of nature. They denigrated ideas that
they found objectionable or incompatible with their "age of reason,"
calling them ignorant or superstitious. Nevertheless, the world
inhabited by these enlightened thinkers was as filled with enchantment
as it had been for people living in the fifteenth and sixteenth centuries.
What people had once called "magic," the Enlightenment called
"science."

The triumphal narrative of the Enlightenment, written first by "the
enlightened" themselves and then taken up by those who came after
them, depicted a glorious new world ruled by reason and liberty, free
from the tyranny of ignorance, superstition, and mindless tradition.
This rhetoric, all but overflowing with a shining kind of idealism, is
compelling even now – it seems familiar to many of us today, perhaps
because we still find traces of these ideals in many of the institutions of
the modern West. Ultimately, though, the Enlightenment was more
complicated and contradictory than this narrative suggests. Its proponents and supporters tried to make a new world, and in some ways they
succeeded. In other ways, however, they did not. Not unlike the
natural philosophers who tried to overthrow Aristotle in the seventeenth century but whose worldviews were shaped irrevocably by the

[1] Sir David Brewster, *Memoirs of the Life, Writings and Discoveries of Sir Isaac
Newton* (Edinburgh, 1855), vol. 2, pp. 374–5.

very thing they wanted to dismantle, the great thinkers and reformers
of the Enlightenment never quite escaped the society they wanted to
transform. True liberty and freedom were still reserved for the elite
few, while the pursuit of "reason" justified ideas that were decidedly
irrational.

Conclusion

Whatever the successes and failures of the great project that was the Enlightenment, it is worth looking back over the preceding centuries to remind ourselves how radically the world changed for European people. *The influence of classical antiquity* was one of the major themes of this book, and in the earliest chapters we saw how firmly the European gaze was fixed on the distant past. By the middle of the eighteenth century, Europe was in the midst of another cultural movement defined instead by a gaze directed to the horizon ahead. Nevertheless, the ancient world has never lost its hold on the Western imagination, at least not entirely. From the eighteenth century there have been periodic revivals of classical themes in architecture, art, philosophy, and literature, and to this day millions of people admire pieces of classical statuary in museums and galleries or visit sites like the Acropolis of Athens and the Roman Colosseum.

Where the influence of antiquity has waned is in our collective understanding of the natural world. The preeminence of Aristotle, Ptolemy, and Galen lasted for almost 2,000 years, but throughout this book we encountered individuals who sought to understand the cosmos in ways that were different from the philosophies of antiquity. In histories of the "Scientific Revolution," men like Copernicus, Paracelsus, Descartes, and Francis Bacon are hailed as reformers and innovators who carved modernity from the solid, weighty philosophies of the past in the same way that the artist Michelangelo (1475–1564) described freeing a sculpture hidden within a block of marble with chisel and mallet. These attempts to abandon the teachings of the ancients were often imperfect or limited, but taken together they represent a crucial shift in the European mindset that paved the way for new ways of studying and understanding the world.

The Enlightenment vision of an informed and educated citizenry drove a series of developments in the eighteenth century that opened up the methods and discoveries of science to larger and larger audiences. A member of the middle classes living in 1750 would have been exposed to mainstream scientific ideas in a way that hardly existed a century earlier. Information was now conveyed in the vernacular rather than in Latin, and natural philosophers recognized an opportunity both to educate the public and to secure sources of financial support and social prestige by staging demonstrations and exhibitions open to everyone, including women and children. More than at any previous point in European history, the average person living in the eighteenth century had opportunities to see and understand the new world described by mathematicians, taxonomists, and geologists.

Another profound shift in how European people understood the world occurred in *the relationship between God and nature*. Every chapter in this book has explored different facets of this relationship, a demonstration of how central it was to the premodern worldview. Some classical philosophies, like Aristotelianism, had virtually no room whatsoever for a deity, while others, like Epicureanism, had as their goal the diminution or rejection of divine causation in the universe. With the widespread acceptance of Christianity in the early centuries of the Common Era, however, large numbers of people started to consider the role of an omnipotent, omniscient God in the natural world. Some ancient philosophies of nature, like that of Plato and his Neoplatonist followers, lent themselves relatively easily to the Christian conception of a singular and all-powerful deity, but European philosophers and theologians in the Middle Ages struggled to reconcile the teachings of pagans such as Aristotle with the foundations of Christian belief and doctrine. The intellectual flourishing of the Renaissance, sparked by the recovery of ideas and texts new to Western Europe, included a deep fascination with the *prisca sapientia*, the ancient wisdom of Creation. Philosophers as disparate as Marsilio Ficino, John Dee, Francesco Patrizi, and Robert Fludd sought to bypass centuries of degeneration and touch the mind of God by reading the Book of Nature in new and powerful ways, guided by those with an older and more perfect understanding.

The chaos of the Reformation and the splintering of Christendom made that task more difficult as there was now widespread and acrimonious disagreement about the very nature of faith. Thus, from the sixteenth century onward we see a shift in how people understood the relationship between God and His creation. Philosophers and naturalists remained as pious as before; consider Johannes Kepler "thinking God's thoughts after Him," or Paracelsus wandering the world in search of the divine secrets hidden in nature. Yet, the religious anxieties that led Dee to converse with angels, that landed Galileo in front of the Inquisition, and that drove both Descartes and Gassendi to demonstrate the presence of God in their mechanical philosophies all suggest that the relationship between God and nature, once assured, was now the subject of question and doubt. When Newton suggested that comets were sent periodically by God to correct imbalances in the vast cosmic machine – yet another attempt to demonstrate God's presence – the German mathematician Gottfried Leibniz (1646–1716) accused Newton of making God seem like an inferior mechanic forced to tinker with an imperfect universe. In Leibniz's outrage we catch a glimpse of a profound anxiety that existed around the turn of the eighteenth century, one motivated by depictions of God as mere caretaker, winding up the cosmic watch and then walking away.

Newton, however, was committed absolutely to the idea that the Creator remained present in His creation, proposing at one point that universal gravitation was the invisible hand of God at work in the cosmos. There is a deep irony, then, in the fact that many philosophers in the eighteenth century interpreted the Newtonian universe as one ruled not by God, but by mathematics and reason. The rise of deism and its distant, unknowable God went hand-in-hand with the proliferation of Newtonian science, thanks in part to efforts by leading Enlightenment thinkers like Voltaire and Diderot to separate organized religion from secular institutions. In response, some theologians and philosophers proposed new evidence for the presence of God. For example, the English clergyman William Paley (1743–1805) published his *Natural Theology: or, Evidences of the Existence and Attributes of the Deity* in 1802 and argued that the presence of design in nature was clear evidence for the existence of God. Paley is known today for his "watchmaker analogy," which claims that the intricate complexity found in many living things must be the result of deliberate design rather than chance or accident, and which remains a central idea held by present-day proponents of creationism and "intelligent design."

For all of these developments, however, the typical European person in the eighteenth century had a religious outlook that was largely unchanged from that held by previous generations. Most Christians went to church each week, followed the teachings of the Bible, and shared an understanding of God that would not have been out of place in the seventeenth century. Popular religious movements such as Pietism or the revivalist fervor of the Great Awakenings were grass-roots affairs, inspired not by sophisticated theologies but by broad social trends and attitudes. In some cases, however, changes to religious attitudes and practices had their roots in the ideas of the educated elite, as in the increasing emphasis on religious tolerance that was encouraged and mandated by Enlightened monarchs and governments.

For most people, then, the unseen hand of God remained present in the universe. They were untroubled by *the problem of occult or hidden causes* that had preoccupied generations of philosophers and theologians. Even in antiquity, Aristotle and Plato had struggled to define not just the role of hidden causes in the universe but also the question of how to study phenomena that could be known only by their effects. The universe bequeathed to the eighteenth century by Isaac Newton solved this problem not by banishing or revealing occult causes, however; on the contrary, he made occult causation central to his philosophy. When Leibniz criticized Newton's explanation for universal gravitation as lacking a clear description of its causes, the latter agreed that his work described "general Laws of Nature" whose "Causes be not yet discover'd." In fact, Newton seemed unconcerned that the causes for gravitation were hidden. His natural philosophy described the effects of gravity on matter – what he called "manifest Qualities" – while conceding that "their Causes...are occult."[1] Thus, Newton resolved the problem of occult or hidden causes by suggesting that it was not a problem at all. Someone could use Newtonian methods to measure and understand gravity's behavior without needing to know anything at all about what caused it.

That Newton was able to sideline or ignore the problem of occult causation owes a significant debt to the mechanical philosophies that had appeared some decades earlier. The tiny atoms of Gassendi or the invisible corpuscles of Descartes were no less occult than the sympathies and correspondences of the hermeticists or the hidden properties

---

[1] Isaac Newton, *Opticks, Based on the Fourth Edition London, 1730* (New York: Dover Publications, 1952), Book III, Part I, Query 31, p. 401.

of the Aristotelians; none of these things were visible to ordinary perception. Yet, there had been relatively little concern from contemporaries that these mechanical causes for natural phenomena were hidden from sight – even if Cartesian corpuscles remained invisible, someone still could infer their motions and behaviors by reference to natural laws and geometrical principles. The widespread acceptance of mechanical explanations for natural phenomena meant that, by the latter decades of the seventeenth century, mainstream philosophies of nature had already embraced occult causes. It was hardly more problematic for Newton to describe the action of an invisible force such as gravity on similarly invisible pieces of matter.

Thus, the Newtonian universe was one in which occult causation was the rule and not the exception. By 1750, most of the European middle classes understood that universe to be a vast expanse in which the Earth was merely one planet among many. What a difference from the small, contained cosmos known to the educated elite of the Middle Ages, which ended just beyond Saturn's orbit at the sphere of the fixed stars. In such a realm, where humanity was both the literal and figurative center of everything, *the interconnectedness of the premodern world* made a deep and intuitive sense to many people. The relationship between microcosm and macrocosm, the practice of sympathetic magic, the influence of the heavens on human health, personalities, and events – all sprang from an understanding of the universe in which everything had its proper and natural place within a complex web of correspondences and associations. By and large, however, Enlightenment philosophers rejected the mystical and spiritual elements of the Renaissance worldview in which humanity, Nature, and God all existed as part of an interconnected whole. What persisted into the eighteenth and nineteenth centuries was a desire to understand humanity's place in the wider universe. Attempts by the French biologist Jean-Baptiste Lamarck (1744–1829) to understand all life in the context of evolutionary change, by Carl Linnaeus to integrate humans into biological taxonomies, and by Georges Cuvier to reconcile human history with geological and paleontological discoveries all suggest that this theme of interconnectedness was transmuted rather than dismantled. Humanity had been displaced from the center of the physical universe by the ideas of Copernicus and Galileo, but metaphorically we humans remained the polestar around which all of Nature revolved.

The Enlightenment was not a blank slate on which Europeans sketched a new world. It was more like a piece of old parchment imperfectly scraped clean, still bearing traces of past ideas around which modernity took shape. The many ways in which premodern people made sense of their universe continue to reverberate into the twenty-first century, linking each of us to other times and worlds, and we are richer for it.

Bibliographical Essays

General Works on Premodern Science, Religion, and Magic

There are many useful surveys of the complex relationship between and among science, religion, and magic in premodern Europe. These include Steven Marrone's *A History of Science, Magic and Belief: From Medieval to Early Modern Europe* (Palgrave Macmillan, 2014), Allison Coudert's *Religion, Magic, and Science in Early Modern Europe and America* (Praeger, 2011), Randall Styers's *Making Magic: Religion, Magic, and Science in the Modern World* (Oxford University Press, 2004), and Stanley J. Tambiah's *Magic, Science, Religion, and the Scope of Rationality* (Cambridge University Press, 1990). For those interested in the specific relationship between science and religion, an important text is John Hedley Brooke's *Science and Religion: Some Historical Perspectives* (Cambridge University Press, 1991); also useful is the subsequent *Science and Religion: New Historical Perspectives*, edited by Thomas Dixon, Geoffrey Cantor, and Stephen Pumfrey (Cambridge University Press, 2010).

Other works examine the connections between science, magic, and religion without drawing attention to the specific ties between them. Examples include Peter Marshall's accessible *The Theatre of the World: Alchemy, Astrology and Magic in Renaissance Prague* (McClelland & Stewart, 2006), as well as the excellent *Wonders and the Order of Nature, 1150–1750* by Lorraine Daston and Katharine Park (Zone Books, 1998). Another useful overview of early modern ideas is provided in *Making Knowledge in Early Modern Europe: Practices, Objects, and Texts, 1400–1800*, edited by Pamela H. Smith and Benjamin Schmidt (University of Chicago Press, 2007).

Background on Premodern Europe

Helpful surveys of European life and society from the Middle Ages through the early modern period include Helmut Georg Koenigsberger's *Medieval Europe, 400–1500* (Routledge, 2014) and Merry E. Wiesner-Hanks's *Early Modern Europe, 1450–1789* (Cambridge University Press, 2006). Closer studies of premodern rural society must include Carlo Ginzburg's seminal *The Cheese and the Worms: The Cosmos of a Sixteenth-Century Miller* (repr. ed.; Johns Hopkins University Press, 2013) as well as Kenneth Jupp's useful survey of feudalism in his article, "European Feudalism from Its Emergence through Its Decline," *American Journal of Economics and Sociology* 59 (2000): 27–45.

Those with an interest in the Black Death and its impacts on premodern Europe should read *King Death: The Black Death and Its Aftermath in Late-Medieval England* by Colin Platt (Routledge, 2014) and John Kelly's *The Great Mortality: An Intimate History of the Black Death* (Harper Perennial, 2006). A classic text documenting the creation of the European university is Charles Haskins's *The Rise of Universities* (repr. ed.; Routledge, 2002); also useful is Alan B. Cobban's *The Medieval Universities: Their Development and Organization* (Taylor & Francis, 1975). Nancy Siraisi's *History, Medicine, and the Traditions of Renaissance Learning* (University of Michigan Press, 2007) presents a more recent examination of learning and teaching in the Renaissance, while those who want to understand the nature and impact of the philosophy of Aristotle must read Charles Schmitt's *Aristotle and the Renaissance* (Harvard University Press, 1983).

On printing and literacy in the Renaissance, see Elizabeth L. Eisenstein's *The Printing Revolution in Early Modern Europe* (Cambridge University Press, 2005), Robert Allan Houston's *Literacy in Early Modern Europe* (Routledge, 2014), and Andrew Pettegree's *The Book in the Renaissance* (Yale University Press, 2010).

Andrew Johnston provides a useful overview of the Reformation in *The Protestant Reformation in Europe* (Routledge, 2016). For more on religion during and after the Reformation, see *Religious Thought in the Reformation* by Bernard Reardon (Routledge, 2014) and *Religion and Superstition in Reformation Europe*, edited by Helen Parish and William G. Naphy (Manchester University Press, 2002).

Chapter 1 Hermeticism, the Cabala, and the Search for Ancient Wisdom

There are many excellent works on the European Renaissance. If you're interested in the translation movement that brought the works of classical antiquity into the Arabic and Islamic worlds, see Dimitri Gutas's *Greek Thought, Arabic Culture: The Graeco-Arabic Translation Movement in Baghdad and Early 'Abbasaid Society (second-fourth/fifth-tenth c.)* (Routledge, 2012) as well as Azzedine Haddour's "Tradition, Translation and Colonization: The Graeco-Arabic Translation Movement and

Deconstructing the Classics," in *Translation and the Classic: Identity as Change in the History of Culture*, edited by Alexandra Lianeri and Vanda Zajko (Oxford University Press, 2008).

The *Renaissance in National Context*, edited by Roy Porter and Mikulas Teich (Cambridge University Press, 1992), Lisa Jardine's *Worldly Goods: A New History of the Renaissance* (W. W. Norton & Company, 1998), and Bard Thompson's *Humanists and Reformers: A History of the Renaissance and Reformation* (William B. Eerdmans Publishing, 1996) all provide a broad perspective on the Renaissance. Works that focus more specifically on Renaissance humanism include Tony Davies's *Humanism* (Routledge, 2008), Thomas DaCosta Kaufmann's *The Mastery of Nature: Aspects of Art, Science, and Humanism in the Renaissance* (Princeton University Press, 1993), and *The Cambridge Companion to Renaissance Humanism*, edited by Jill Kraye (Cambridge University Press, 1996).

For a broad introduction to the history of magic in premodern Europe, an excellent resource is Richard Kieckhefer's *Magic in the Middle Ages* (2nd ed.; Cambridge University Press, 2014). See also Euan Cameron's *Enchanted Europe: Superstition, Reason, and Religion, 1250–1750* (Oxford University Press, 2011) and Valerie Flint's *The Rise of Magic in Early Medieval Europe* (Princeton University Press, 1991). For more information about folk or popular magic, a useful study is Owen Davies's *Popular Magic: Cunning-Folk in English History* (A&C Black, 2007).

Those wishing to learn more about ritual or learned magic should read Christopher Lehrich's *The Occult Mind: Magic in Theory and Practice* (Cornell University Press, 2007), Owen Davies's *Grimoires: A History of Magic Books* (Oxford University Press, 2010), and Benedek Láng's *Unlocked Books: Manuscripts of Learned Magic in the Medieval Libraries of Central Europe* (Penn State, 2010). On a related note, the fascinating ideas of magical practitioners can be found in works like Deborah Harkness's *John Dee's Conversations with Angels: Cabala, Alchemy, and the End of Nature* (Cambridge University Press, 1999) and William Huffman's *Robert Fludd and the End of the Renaissance* (Routledge, 1988). One of the most significant studies of early modern hermeticism remains Frances Yates's *Giordano Bruno and the Hermetic Tradition* (University of Chicago Press, 1991). For more recent work on the Hermetic tradition, see Florian Ebeling's *The Secret History of Hermes Trismegistus: Hermeticism from Ancient to Modern Times* (Cornell University Press, 2007). On Ficino's particular interpretation of hermeticism, see Denis Robichaud's "Ficino on Force, Magic, and Prayers: Neoplatonic and Hermetic Influences in Ficino's Three Books on Life," *Renaissance Quarterly* 70 (2017): 44–87.

On Renaissance cabalism, see Brian Copenhaver's "The Secret of Pico's *Oration*: Cabala and Renaissance Philosophy," *Midwest Studies in*

Philosophy 26 (2002): 56–81. Also useful is Philip Beitchman's *Alchemy of the Word: Cabala of the Renaissance* (SUNY, 1998).

Chapter 2 Witchcraft and Demonology

Those who want to learn more about the Cathars should read Mark Gregory Pegg's *A Most Holy War: The Albigensian Crusade and the Battle for Christendom* (Oxford University Press, 2008) and Malcolm D. Lambert's *The Cathars* (Blackwell, 1998). More general information about premodern European heresy is available in Lambert's *Medieval Heresy: Popular Movements from the Gregorian Reform to the Reformation* (Blackwell, 1992), while John Tedeschi examines the role of the Inquisition in the pursuit of heretics in his *The Prosecution of Heresy: Collected Studies on the Inquisition in Early Modern Italy* (Medieval & Renaissance Texts & Studies, 1991).

There are many sources that describe various aspects of the European witch hunts. One of the most helpful and comprehensive studies is Brian P. Levack's *The Witch-Hunt in Early Modern Europe* (4th ed.; Routledge, 2015); other broad-ranging works include *Witchcraft in Early Modern Europe: Studies in Culture and Belief*, edited by Jonathan Barry, Marianne Hester, and Gareth Roberts (Cambridge University Press, 1998), Gary K. Waite's *Heresy, Magic, and Witchcraft in Early Modern Europe* (Palgrave Macmillan, 2003), Charles Zika's *Exorcising Our Demons: Magic, Witchcraft, and Visual Culture in Early Modern Europe* (Brill, 2003), and Michael D. Bailey's *Battling Demons: Witchcraft, Heresy, and Reform in the Late Middle Ages* (Penn State, 2003). Alan Macfarlane's *Witchcraft in Tudor and Stuart England: A Regional and Comparative Study* (2nd ed.; Routledge, 2008) remains an excellent examination of early modern witchcraft at a more focused and regional level.

On early modern demonology, Stuart Clark's *Thinking with Demons: The Idea of Witchcraft in Early Modern Europe* (Oxford University Press, 1997) is a superb (if daunting) resource. Also helpful is Jan Machielsen's "Thinking with Montaigne: Evidence, Scepticism, and Meaning in Early Modern Demonology," *French History* 25 (2011): 427–52. Walter Stephens's *Demon Lovers: Witchcraft, Sex, and the Crisis of Belief* (University of Chicago Press, 2002) does an excellent job of both explaining and problematizing early modern ideas about demons and witches.

Chapter 3 Magic, Medicine, and the Microcosm

For broad surveys of medicine and medical practices in premodern Europe, see *Early Modern Medicine and Natural Philosophy*, edited by Peter

Distelzweig, Benjamin Goldberg, and Evan R. Ragland (Springer, 2016); Mary Lindemann's *Medicine and Society in Early Modern Europe* (Cambridge University Press, 2010); and Nancy Siraisi's *Medieval and Early Renaissance Medicine* (University of Chicago Press, 1990).

Useful information about early modern astrology can be found in Nicholas Campion's *A History of Western Astrology, Volume II: The Medieval and Modern Worlds* (Bloomsbury, 2009). Anthony Grafton's *Cardano's Cosmos: The Worlds and Works of a Renaissance Astrologer* (Harvard University Press, 1999) is also essential reading, especially for its discussions of the links between astrology and medicine.

An accessible introduction to the figure of Paracelsus can be found in Philip Ball's *The Devil's Doctor: Paracelsus and the World of Renaissance Magic and Science* (Farrar, Straus and Giroux, 2006). More comprehensive is the collection of essays in *Paracelsus* (Brill, 1998), edited by Ole Peter Grell. Walter Pagel's *Paracelsus: An Introduction to Philosophical Medicine in the Era of the Renaissance* (Karger Medical and Scientific Publishers, 1982) remains an important study of Paracelsus; also worth reading is Pagel's biography of the Paracelsian physician Van Helmont: *Joan Baptista van Helmont: Reformer of Science and Medicine* (Cambridge University Press, 1982).

Finally, those wishing to learn more about the weapon salve or the powder of sympathy should read Elizabeth Hedrick's "Romancing the Salve: Sir Kenelm Digby and the Powder of Sympathy," *The British Journal for the History of Science* 41 (2008): 161–85, as well as my own "The Perversion of Nature: Johannes Baptista van Helmont, the Society of Jesus, and the Magnetic Cure of Wounds," *Canadian Journal of History* 38 (2003): 179–97.

Chapter 4 A New Cosmos: Copernicus, Galileo, and the Motion of the Earth

Two important studies on premodern astronomy and the changes it experienced in early modern Europe are Arthur Koestler's *The Sleepwalkers: A History of Man's Changing Vision of the Universe* (Penguin Books, 1990) and Thomas Kuhn's *The Copernican Revolution: Planetary Astronomy in the Development of Western Thought* (Harvard University Press, 1992). Edward Grant's *Planets, Stars, and Orbs: The Medieval Cosmos, 1200–1687* (Cambridge University Press, 1994) also does an excellent job of explaining the universe as understood by early modern Europeans. For those who wish to grapple with the practice of "saving the phenomena," see Pierre Duhem, *To Save the Phenomena: An Essay on the*

Idea of Physical Theory from Plato to Galileo, translated by Edmund Dolan and Chaninah Maschler (University of Chicago Press, 1969). On the important connections between religion and astronomy, see John L. Heilbron's *The Sun in the Church: Cathedrals as Solar Observatories* (Harvard University Press, 1999).

For more on Copernicus and his revolution, see Robert S. Westman's *The Copernican Question: Prognostication, Skepticism, and Celestial Order* (University of California Press, 2011); Owen Gingerich's *The Book Nobody Read: Chasing the Revolutions of Nicolaus Copernicus* (Bloomsbury, 2004); and Dennis Richard Danielson's *The First Copernican: Georg Joachim Rheticus and the Rise of the Copernican Revolution* (Bloomsbury, 2006).

There are many studies of Galileo and his infamous encounter with the Catholic Church. John L. Heilbron's *Galileo* (Oxford University Press, 2012) provides a useful and fairly recent overview. Of particular note as well are several works written by Maurice Finocchiaro; important examples include *Defending Copernicus and Galileo: Critical Reasoning in the Two Affairs* (Springer, 2010) and *Retrying Galileo, 1633–1992* (University of California Press, 2005). Rivka Feldhay's *Galileo and the Church: Political Inquisition or Critical Dialogue?* (Cambridge University Press, 1995) does an excellent job of highlighting the politics within the Church that influenced the outcome of Galileo's trial.

Chapter 5 Looking for God in the Cosmic Machine

For an accessible overview of the mechanical philosophies, see Richard S. Westfall's *The Construction of Modern Science: Mechanisms and Mechanics* (John Wiley, 1971). Also important is Margaret J. Osler's *Divine Will and the Mechanical Philosophy: Gassendi and Descartes on Contingency and Necessity in the Created World* (Cambridge University Press, 1994), which explores the connections between theology and the mechanical philosophies of Gassendi and Descartes. For a more recent discussion of the early mechanical philosophies, see Klaas Van Berkel's *Isaac Beeckman on Matter and Motion: Mechanical Philosophy in the Making* (Johns Hopkins, 2013). On the mechanical philosophies as they existed after Descartes and Gassendi, see Thomas M. Lennon's *The Battle of the Gods and Giants: The Legacies of Descartes and Gassendi, 1655–1715* (Princeton University Press, 2014). Also useful, especially in the context of early modern debates about vitalism, is Geert Jan M. De Klerk, "Mechanism and Vitalism. A History of the Controversy," in *Acta Biotheoretica* 28 (1979): 1–10.

On Gassendi, see Barry Brundell's *Pierre Gassendi: From Aristotelianism to a New Natural Philosophy* (Springer, 2012); Antonia LoLordo's *Pierre Gassendi and the Birth of Early Modern Philosophy* (Cambridge University Press, 2006); and Saul Fisher's *Pierre Gassendi's Philosophy and Science: Atomism for Empiricists* (Brill, 2005).

On Descartes, a useful and recent overview is provided by Georges Dicker in *Descartes: An Analytic and Historical Introduction* (Oxford University Press, 2013). Also important in understanding Cartesian natural philosophy is Joseph Frederick Scott's *The Scientific Work of René Descartes: 1596–1650* (Routledge, 2016) and Stephen Gaukroger's *Descartes' System of Natural Philosophy* (Cambridge University Press, 2002). The philosopher Dennis Des Chene has also placed some of Descartes's ideas in the context of important philosophical debates of the time; see his *Spirits and Clocks: Machine and Organism in Descartes* (Cornell University Press, 2001) and *Physiologia: Natural Philosophy in Late Aristotelian and Cartesian Thought* (Cornell University Press, 1996).

Finally, on the encounter between Joseph Glanvill and John Webster, see Thomas H. Jobe, "The Devil in Restoration Science: The Glanvill-Webster Witchcraft Debate," in *Isis* 72 (1981): 342–56.

Chapter 6 Manipulating Nature: Experiment and Alchemy in the Scientific Revolution

On experience and experiment in early modern Europe, see *The Science of Nature in the Seventeenth Century: Patterns of Change in Early Modern Natural Philosophy* (Springer, 2006), edited by Peter R. Anstey and John A. Schuster; *Historia: Empiricism and Erudition in Early Modern Europe* (MIT, 2005), edited by Gianna Pomata and Nancy Siraisi; Steven Shapin and Simon Schaffer's *Leviathan and the Air Pump: Hobbes, Boyle, and the Experimental Life* (Princeton University Press, 1989); Steven Shapin's *A Social History of Truth: Science and Civility in Seventeenth-Century England* (University of Chicago Press, 1994); Peter Dear's *Discipline and Experience: The Mathematical Way in the Scientific Revolution* (University of Chicago Press, 1995); Lorraine Daston, "The Nature of Nature in Early Modern Europe," in *Configurations* 6 (1998): 149–72; and Peter Dear, "Miracles, Experiments, and the Ordinary Course of Nature," in *Isis* 81 (1990): 663–83. Those wishing to read the ideas of Francis Bacon can do so in a recent edition of two of his most important works: *New Atlantis and The Great Instauration* (Wiley-Blackwell, 2016), edited by Jerry Weinberger. On the history of alchemy, an excellent overview is provided by Lawrence M. Principe in his *The Secrets of Alchemy* (University of Chicago Press,

2012). Principe has also demonstrated the central role played by alchemy in the ideas of Robert Boyle; see his *The Aspiring Adept: Robert Boyle and His Alchemical Quest* (Princeton University Press, 2000). William Newman has also written several important studies of early modern alchemy; two examples are *Atoms and Alchemy: Chymistry and the Experimental Origins of the Scientific Revolution* (University of Chicago Press, 2006) and *Promethean Ambitions: Alchemy and the Quest to Perfect Nature* (University of Chicago Press, 2004). Also useful are works by Bruce Moran: *Andreas Libavius and the Transformation of Alchemy: Separating Chemical Cultures with Polemical Fire* (Science History Publications, 2007) and *Distilling Knowledge: Alchemy, Chemistry, and the Scientific Revolution* (Harvard University Press, 2005). To learn more about the female alchemist Anna Zieglerin, see Tara Nummedal, "Alchemical Reproduction and the Career of Anna Maria Zieglerin," in *Ambix: The Journal of the Society for the History of Alchemy and Early Chemistry* 49 (2001): 56–68. For more on the decline of alchemy around the turn of the eighteenth century, see Marieke M. A. Hendriksen, "Criticizing Chrysopoeia? Alchemy, Chemistry, Academics, and Satire in the Northern Netherlands, 1650–1750," in *Isis* 109 (2018): 235–53.

Chapter 7 A New World? The Dawn of the Enlightenment

For an excellent overview of the Enlightenment, see Dorinda Outram's *The Enlightenment* (3rd ed.; Cambridge University Press, 2013); another useful collection is *The Enlightenment World* (Routledge, 2004), edited by Martin Fitzpatrick, Peter Jones, Christa Knellwolf, and Iain McCalman. On the development of public science in the Enlightenment, see *Science and Spectacle in the European Enlightenment* (Routledge, 2016), edited by Bernadette Bensaude-Vincent and Christine Blondel; on the problematic uses of science in the eighteenth century, see Londa Schiebinger's *Nature's Body: Gender in the Making of Modern Science* (2nd ed.; Rutgers, 2004). S. J. Barnett's *The Enlightenment and Religion: The Myths of Modernity* (Manchester, 2003) and James M. Byrne's *Religion and the Enlightenment: From Descartes to Kant* (Westminster John Knox Press, 1997) both provide helpful information about religion in the eighteenth century. Keith Thomas's *Religion and the Decline of Magic* (Scribner, 1971) remains a useful survey of popular ideas about magic, especially when complemented by John Henry's excellent "The Fragmentation of Renaissance Occultism and the Decline of Magic," in *History of Science* 46 (2008): 1–48.

Index

Abassid dynasty, the, 19
Académie Royale des Sciences, the, 192
Accademia del Cimento, the, 192
Al-Bīrūnī, 20
alchemy, 9, 12, 30, 75, 83, 94, 161, 174–83, 189–90, 200
  and replication, 181
  and secrecy, 179, 181
  chrysopoeia, 176, 189, 200
  Philosopher's Stone, the, 177–83
alchemy, 43
Al-Farabi, 20
Al-Kindī. *See* Alkindus
Alkindus, 20
Al-Rāzī. *See* Rhazes
Aquinas, St. Thomas, 31
Aristotelian philosophy, 35, 104–7, 136, 161, 167–9, 184, 204, 207
  and experience, 165–7
  cosmology, 105–7, 111, 125
  on the soul, 150, 153
  theory of metals, 176–7
Aristotle, 20, 81, 104–8, 111, 113, 125, 130, 132, 138, 150–1, 162–6, 168, 176, 201, 203–4, 206
astrology, 10, 29, 40, 75, 85–90, 207
  horoscopes, 85–6
astronomy, 85, 102–34
  Copernican system, the, 113–15, 127, 130
  Ptolemaic system, the, 108–11, 113, 127, 130
  saving the phenomena, 108–10
  Tychonic system, the, 119–20
atheism, 9, 151, 158, 197
atomism, 138–42, 144, 155
Averroes, 20
Avicenna, 20, 76, 80, 93, 97

Bacon, Francis, 11, 165–9, 174, 183, 203
  and inductive reasoning, 166–9, 184
Báthory, Stefan, King of Poland, 42
Bayt al-Hikmah. See House of Wisdom, the
Bellarmine, St. Robert, 130–1
Black Death, the, 8, 47
Bodin, Jean, 59
Boerhaave, Herman, 189, 200
Boyle, Robert, 11, 171–4, 182–3, 189
Bracciolini, Poggio, 141
Brahe, Tycho, 119–20, 122
Brewster, Sir David, 200
Bruno, Giordano, 36
Buffon, Comte de, 194
Burnet, Gilbert, 182
Byzantine Empire, the, 17–19

cabalism, 13, 23, 33–6, 39–41, 49, 95, 97, 199
Cartesian philosophy, 143–9
  laws of motion, 144–6
  on the soul, 151–4
  physics and cosmology, 143–6
  res cogitans, 144, 152, 154–5
  res extensa, 144, 152, 154–5
Casaubon, Isaac, 27–8
Castelli, Benedetto, 128
Catharism. *See* heresy: Catharism
Catholic Church, the, 2, 11, 22, 32, 47, 49, 73, 99–100, 103–4, 128–32, 191
Cavendish, Margaret, Duchess of Newcastle-upon-Tyne, 155
Celsus, 91
chymistry, 175–6, 183, 189–90, 200
Cicero, 22

classical antiquity, 9–10, 16–19, 21–3, 25, 29, 39, 83, 91, 96, 111, 114, 138, 166, 203, 206
conflict between science and religion, 103–4, 132–4
Constantine the Great, 17
Copernicus, Nicolaus, 11, 83, 112–19, 121–2, 127, 161, 183, 203, 207
    De revolutionibus, 112–18
Cremonini, Cesare, 124
Cuvier, Georges, 194, 207

Darwin, Charles, 194
De' Medici, Cosimo I, 13, 20, 122
De' Medici, Cosimo II, 127–8, 172
Dee, John, 14, 37, 39–43, 58, 63, 95–7, 122, 199, 204
deism, 197, 205
Della Porta, Giambattista, 97, 159
Democritus, 104
Descartes, René, 11, 135, 142, 154, 156, 171, 184, 193, 197, 199, 203, 205–6
    and deductive reasoning, 166
    Meditations on First Philosophy, 147–9
Devil, the, 37, 61–7, 71, 99, 157
Diderot, Denis, 195, 205

Einstein, Albert, 183
empiricism, 11, 101, 142, 148, 158, 160–1, 168, 198
Enlightenment, the, 10, 12, 23, 138, 160, 169, 175, 187, 189–201, 203, 207–8
    and magic, 198–201
    and religion, 195–8
    and science, 192–5, 204
Epicurus, 104, 138–42, 150
Erasmus, 22
experimentation, 11, 157–8, 160–1, 163–5, 168, 175
    and credibility, 169–74, 179
    and replication, 170, 173, 179

Faustus, 7, 14, 37–9, 58, 63
feudalism, 7, 47
Ficino, Marsilio, 13–16, 20, 25, 27–8, 31–2, 40–1, 96, 122, 204
Fludd, Robert, 32, 99, 204

Forman, Simon, 89
Foster, William, 99
Fraternity of the Rosy Cross. *See* Rosicrucians
Frisius, Paulus, 62

Galen, 76, 81, 92, 95, 203
Galilei, Galileo, 11, 40, 103–4, 108, 120, 122–35, 143, 148, 161, 171, 184, 205, 207
    Dialogue Concerning the Two Chief World Systems, 130–2
    Letter to the Grand Duchess Christina, 128–9
    Siderius nuncius, 125–7
Gassendi, Pierre, 11, 135, 138–42, 144, 146, 148–51, 154, 156, 184, 197, 199, 205–6
    on the soul, 150–1
geniture. *See* astrology: horoscopes
geocentrism, 11, 85, 104–11
Giorgi, Francesco, 36
Glanvill, Joseph, 158–60
Goclenius the Younger, Rudolph, 100–1
Great Awakenings, the, 196, 206
Grillandi, Paolo, 62
Guazzo, Francesco Maria, 59

Hegel, Georg Wilhelm Friedrich, 195
heliocentrism, 113, 120, 143, 184, 187
Hellenistic period, the, 17, 21, 26, 98
Helmont. *See* Van Helmont
heresy, 10, 47–9, 52, 59–60, 73, 100, 131
    Catharism, 48–9, 66
Hermes Trismegistus, 13, 16, 24–5, 29
hermeticism, 10, 13, 16, 23, 27–33, 35–6, 39–42, 49, 51, 84, 94–5, 97, 199, 206
    Corpus Hermeticum, the, 13, 15–16, 29–31, 41, 122
    Pimander, the, 29
Hippocrates, 76–8, 86
Horace, 190
House of Wisdom, the, 19
humanism, 16, 21–3, 36, 81, 96, 138, 147

Iamblichus, 22, 25
Ibn Rushd. *See* Averroes
Ibn Sīnā. *See* Avicenna, *See* Avicenna
Innocent III, Pope, 48
Innocent VIII, Pope, 35, 60
Inquisition, the, 49, 56, 71, 100,
    129–30, 132, 205
Islamic Renaissance, the, 16, 19–20,
    104

John Paul II, Pope, 104

Kabbalah, the, 10, 24, 33–6
    sefirot, the, 34
Kant, Immanuel, 190–1, 195
Kelley, Edward, 42–3
Kepler, Johannes, 113, 120–2, 148,
    161, 184, 205
Keynes, John Maynard, 183
Kircher, Athanasius, 32, 36, 183
Kramer, Heinrich, 60–1

Lamarck, Jean-Baptiste, 207
Leclerc, Georges-Louis. *See* Buffon,
    Comte de
Leibniz, Gottfried, 11, 205–6
Linnaeus, Carl, 194, 207
Lippershey, Hans, 124
Locke, John, 11
Lucretius, 140–1, 150–1
Luther, Martin, 94, 118
Lyell, Charles, 194

magic, 6–7, 10–11, 28, 30, 36, 76, 94
    astrological magic, 30, 88–90, 101
    demonic magic, 6–7, 15, 38–9, 58,
        182
    folk magic, 13, 49–52, 57, 75
    natural magic, 6, 16
    necromancy, 15–16
    sympathetic magic, 98, 207
Maier, Michael, 180
Malleus maleficarum, the. *See*
    witchcraft theory: *Malleus
    maleficarum*, the
Marlowe, Christopher, 37–9
materialism, 139–40, 149, 155, 186,
    197
mathematics, 40, 107, 115–18, 124,
    148–9, 187, 192, 194, 201, 205

mechanical philosophies of nature, the,
    11, 101, 135–61, 185–6, 197,
    205–7
Medici. *See* De' Medici
medicine, 10, 75–101, 164, 189
    apothecaries, 78
    astrological medicine, 86–9
    barber-surgeons, 78–9, 86, 164
    chemical medicine, 95, 100, 159
    folk medicine, 51, 57, 78, 97–8
    humoral system, the, 77–8, 86–8, 92
    physicians, 78–81, 83, 86, 164
    quackery, 89
Michelangelo, 203
microcosm/macrocosm theory, the, 75,
    83–90, 207
miracles, 37, 62–3, 100, 137–8, 156
Muhammad, 19

Napier, Richard, 89
natural philosophy, 4–5
natural theology, 197, 205
Neoplatonism, 22, 35, 40, 96, 204
new science, the, 11, 101, 157–9,
    161–2, 169, 175, 183, 190
Newton, Isaac, 11, 183–7, 194–5, 199,
    205–7
    and alchemy, 184–6, 200–1
Nider, Johannes, 59

Osiander, Andreas, 115–18

paganism, 22, 29, 31, 204
Paley, William, 205
Paracelsus, 11, 75, 78, 90–7, 177, 189,
    199, 203, 205
    doctrine of signatures, the, 94
    tria prima, the, 92–3, 176
Patrizi, Francesco, 31–3, 204
Paul III, Pope, 115, 117
Paul V, Pope, 130
Petrarch, 22
Pico della Mirandola, Giovanni, 31–2,
    35–6
Pietism, 196, 206
Plato, 4, 13, 20, 22, 25, 31, 104, 120,
    204, 206
Pliny the Elder, 98
Plotinus, 22, 25
Postel, Guillaume, 36

powder of sympathy, the, 76, 100–1
prisca sapientia, 24–6, 31, 33, 35–6,
    204
prisca theologia, 24–6, 29, 33, 35–6,
    96, 122
Protestant Reformation, the, 2, 5, 32,
    42, 47, 58, 67, 73, 90, 195, 204
Protestantism, 5, 94–5, 122, 159, 196,
    198
Ptolemy, Claudius, 108–11, 125, 130,
    132, 203

Rashidun Caliphate, the, 19
religion, 4–5, 36, 45, 97
Remy, Nicolas, 59
Renaissance, the, 15–21, 36, 78, 147,
    204, 207
republic of letters, 174
Reuchlin, Johann, 36
Rhazes, 20
Rheticus, 118
Roberti, Jean, 100
Roman Empire, the, 17–19, 25, 27–8,
    81, 98, 104, 141
Rosicrucians, the, 41–2, 99
Rousseau, Jean-Jacques, 193
Royal Society of London, the, 158, 166,
    169, 174, 183, 192
Rudolph II, Holy Roman Emperor,
    42–3, 120

Scientific Revolution, the, 1, 161, 183,
    203
skepticism, 9, 146–7, 158–9
soul, the, 11, 149–57
Sprenger, Jacob, 60
Stahl, Georg Ernst, 155

theology, 4, 24, 32
Titian, 81

Umayyad Caliphate, the, 19–20
universities, 4, 7–8, 15, 21, 124, 165
Urban VIII, Pope, 130–1
Ussher, James, 194

Van Helmont, Jan Baptista, 99–101
Vesalius, Andreas, 80–3, 90, 114
    Fabrica, the, 81–3, 90
vitalism, 154–6, 185
Voltaire, 195, 205

weapon salve, the, 76, 97–101, 159,
    199
Webster, John, 159–60
witch, the, 52–8
    preponderance of women among the
        accused, 55, 57, 65
    stereotypes of accused witches, 52–5
witchcraft, 44–74, 76, 99–100, 157–9
    and Devil-worship, 45, 60, 64–6, 68,
        72
    maleficia, 45, 52–3, 56, 69, 71
witchcraft theory, 52, 58–9, 72
    and the diabolical pact, 61–9
    Malleus maleficarum, the, 60–1, 65
    sabbats, 59, 65–6, 68–70
    witchcraft as sexual crime, 60–1
    witches as heretics, 60, 65
witch-hunts, the, 2, 7–8, 10, 44–7,
    69–74
    accusations against the elderly,
        56–7
    and social dynamics, 55–8
    chain-reaction hunts, 71
    decline and end of the hunts, 72–4
    regional and geographical
        differences, 53–4, 67–73
    use of torture, 44, 69–73

Zieglerin, Anna, 177–8

Printed in the United States
By Bookmasters